# The Machine That Floats

*by*

Joe Gibson

# The Machine That Floats
## by Joe Gibson

Copyright © 2023

All Rights reserved.

No part of this publication may be reproduced, stored in a retrieval system, or transmitted in any form or by any means, electronic, mechanical, photocopying or Otherwise, without the written permission of the publisher.
The author/editor asserts the moral right to be identified as the author/editor of this work.

ISBN: 978-93-59324-27-2

Published by

**DOUBLE 9 BOOKS**

2/13-B, Ansari Road
Daryaganj, New Delhi – 110002
info@double9books.com
www.double9books.com
Tel. 011-40042856

# ABOUT THE AUTHOR

Joe Gibson is a talented and diverse author noted for his ability to create intriguing stories in a variety of genres. Their writing helps individuals connect with and understand one another. Gibson has grabbed readers with his rich imagination and fascinating plots, and his passion for writing shines through in his work. He regularly offers captivating narratives that keep readers anxiously turning the pages, whether going into the realms of science fiction, mystery, fantasy, or beyond. Gibson's writing is distinguished by rich character development, complex story twists, and a keen knowledge of human emotions and relationships. His skill as a storyteller helps readers to thoroughly immerse themselves in the worlds he creates, making each reading experience a riveting journey. Joe Gibson has a unique ability as an author to explore a wide range of themes and genres, demonstrating his versatility and attention to the craft of writing. His books have left an indelible mark on the literary scene, earning him a devoted fanbase as well as critical recognition.

Bill Morrow fished his cigarettes out, shook one loose, and poked it between his lips. He lighted it with hands that shook badly, he leaned back on the workbench and blew smoke in a long, heavy sigh.

His gaze remained fixed on the compact little chunk of glittering grids, coils, and metal loops that floated in the center of the room. Floated, by Isaac Newton—*floated!*

It worked. It worked beautifully! He'd merely inserted the four dry-cell flashlight batteries into their clamps and thumbed the switch on the little face-panel. The tiny pilot-light winked on, the needle jiggled on the single instrument dial—

And it worked. It had risen gently from the workbench, floating into the air....

Then, seemingly, it had fostered a dislike for the workbench. It slid off and bounced toward the floor—bounced, up and down in the air, gently—and floated on across the cellar toward the oil furnace in the corner.

But as it approached the oil furnace, it had decided it didn't like that either—so it deflected its course and floated toward the concrete cellar wall.

But it didn't like the wall. So it reversed its course and retreated to the center of the room. There it hovered, four feet above the cement floor, four feet below the rafters of the cellar roof.

It hovered in mid-air.

Morrow stared at it, critically. He could capture it—get it between himself and the wall, and reach out and grab it before it could slip away—and touching it wouldn't harm him. The magneto-gravitic coils didn't need high voltage.

It was working on its lowest "volume" setting. The only word applicable was "volume" because he used an ordinary volume-control grid and knob to adjust its power—and, again, "power" was the only applicable word. He might have to invent a few new words for it.

But on its lowest volume setting, it was supporting its own weight—suspending itself in the Earth's gravitic field.

And since gravitic forces were also magnetic forces, he would weigh a fraction of a pound lighter when he grabbed hold of the mechanism—just he, himself, since he wore rubber-soled shoes. If he turned up its volume, it would exert greater influence on the molecular structure of itself and of his body—and perhaps of a few grains of dust on the cement floor beneath his feet—by simple mass-attraction and conductivity.

Of course, "mass-attraction" and "conductivity" were also obsolete terms—except that they described two different results of the same natural phenomenon. The floating mechanism affected the basic phenomenon itself—

And

$$_3^i \underline{K};\ \lambda = 0;\ \overline{\Gamma i} = 0;\ Rik = 0;\ \rho,\, \overset{is}{\underset{s}{v}} = 0$$

was the closest Einstein could come to explaining that!

Still, a word could be invented for it, Morrow supposed. Not that he understood what the new word was supposed to define—but then, had Edison known what electricity was? No! He had merely experimented and learned what it would do, and then designed mechanisms which would utilize it.

Morrow didn't know what "gravity and magnetic moment" was, either—nor "angular momentum"—but he had discovered what it would do. *It*, not they—it was all the same thing. And he designed a mechanism. And the mechanism worked.

It defied "gravity."

With its volume turned up, it could very probably lift him to any height above the Earth he desired, with its ability growing weaker only as it rose out of the Earth's gravity and magnetic field. And it would keep him suspended, if he desired, until its batteries burned out.

There would be limitations, of course. Perhaps the Earth's gravity and magnetic fields would be too weak at, say, an altitude of fifty miles for the mechanism to function. There were probably limits to the mass and weight it could lift. There would have to be extensive tests—

And a cellar workshop was no place to conduct them!

He straightened up from the workbench and moved forward on the balls of his feet. He spread his arms wide as he approached the mechanism, like a basketball player approaching a wary opponent who had the ball. Smoke from the cigarette dangling out of the corner of his mouth streamed up and stung his eye. He wished he had left it back on the workbench.

At first, the little mechanism ignored him. Then, almost instinctively, it seemed to notice him. It went sliding away from him, toward the wall.

Morrow moved forward, cautiously.

It glided close to the wall, then rebounded gently. It came drifting back toward him—then hesitated, started off in a tangent—and he grabbed for it. A faint, tingling shock went up his arm as he clawed at the shiny metal loops, but that was all. He hung on grimly as it tugged at his fingertips; then, as its influence swept through him and attuned his body to it, it snuggled up to him, suddenly friendly.

He snapped it off and felt its inert weight settle down familiarly in his hands. He carried it back to the workbench, set it down, and threw a rag over it.

Then he pulled off his coveralls, went upstairs to the kitchenette, and washed his hands.

There were other factors to consider, of course. Especially the ones he didn't want to think about—the frightening ones—

He stared down at his hands, feeling the cool water run pleasantly over them. Strong, supple hands. Well-proportioned, muscular. A little bit like the rest of him. Not fat or skinny, not soft-muscled nor, again, as bulgingly muscular as a wrestler. Just firm flesh, strong and not too much of it, on a strong-boned skeleton frame. Nerves well-coordinated, reflexes good. But tired. Mentally fatigued, the psycho-therapists said, from living in a world of raw tensions. According to them, ninety percent of the American public suffered mental fatigue. There had been a slew of magazine articles and several books about it.

The Cold War, the war that wasn't a war. The Russkies.

Morrow turned off the cold-water tap and glanced at his image in the shaving mirror. A slender face, a good nose, a firm mouth with slightly too much jaw. Dark hair tumbled in comfortable looseness over a lined forehead. Gray eyes that mocked him as he mocked himself.

He dried his hands and got a couple of cans of beer out of the refrigerator. Grabbing a can-opener and a glass, he strolled in through the small, dark bedroom to the front living room and sprawled himself out in the deep chair beside the television set. It was a small home, a comfortable home, and he enjoyed prowling around in it in his socks, loafers, and shorts. He scratched his left leg and opened a can of beer.

He was, Morrow concluded, the product of an age of terror. East was Russia and west was the Allied Nations, and in between was a veritable

No Man's Land. Radar blanketed the skies, rocket missiles stood on their firing-racks, long-range bombers waited to deliver atomic death and swift jet-fighters waited to do battle with them. The diplomats called it a balance of power; the military strategists, a balance of forces, wherein neither side could launch an atomic war without suffering complete annihilation by the other.

And so, said the statesmen, there would be no atomic war.

The only trouble was, they couldn't convince the people. Too many self-minded individuals saw the world situation as two sticks of dynamite rubbing against each other. At any moment, both might explode. Massive war industry and compulsory military training for their youngsters didn't make the public feel any more secure.

Nor, of course, did the generals want them to feel secure. The Allied generals moved their armies in threatening maneuvers near critical borders to increase the fear of the peoples in communist-dominated countries; the Russian generals did likewise to increase the fear of people in the Allied Nations. And the diplomats hurled threats back and forth in the United Nations' assemblies to achieve the same purpose.

Militarily, the two sides had reached a stalemate. The final weapon was the people. Each side hoped the people of the other side would rise up in revolt, thus breaking the deadlock and winning the struggle, but humanity is notoriously stubborn. It was, nonetheless, rather hard on the people.

Individual lives were deeply affected, sometimes for better and sometimes for worse.

Morrow's life had, so far, been for the better. In high school, certain aptitude tests had placed him in advance physics classes; upon graduation, at seventeen, he had spent a year in a government-sponsored engineering school. At eighteen, further tests had placed him in the Air Force, assigned as radar-operator to the rear cockpit of a sleek, all-weather jet-fighter. He spent two years patrolling the stratosphere over the vast, white expanse of the Arctic Ocean. At twenty, he was reassigned to engineering school and spent four years studying electronics, during which time he was returned to civilian status. He was placed at Western Electronics as a production engineer; by the time he was twenty-seven, he had worked his way up to the Research Division. His flying experience helped considerably, but it wasn't all. He was deeply in love with electronics. He had studied Einstein's equations, for example, and got something out of them that most of the others missed. No one knew exactly what it was—neither did Morrow—but he began to have "hunches" that often paid off.

In electronics, that was a priceless faculty. A great deal of it, especially in the research department, was still pretty much of a hit-or-miss affair. There still wasn't a man who knew *exactly* what electricity was!

Now, at twenty-nine, he had gotten another of those "hunches." It worked, too! The machine floated!

He gazed thoughtfully out the broad picture window at the stretch of green lawn, the sidewalks, the trees along the street and the other little prefab houses of his neighbors. The evening shadows were cool and deepening as night approached. Warm, yellow light poured from the windows across the street.

He was comfortable here in his little, company-owned bachelor's home. Most of the town of Westerton was owned by Western Electronics, with its huge, sprawling plant buildings on the other side of the small valley, across the railroad tracks. Like most of the bachelor engineers, he ate most of his meals over at the company cafeteria. A cleaning-woman came twice a week to tidy up his little house, though he was a fairly conscientious housekeeper himself.

And like practically all the engineers, he had a small workshop patched together in his cellar, built from odds and ends salvaged from the company's junk-pile of rejected parts, a few pieces scrounged from the laboratories, and some odd bits made in the machine-shop over the protests of its foreman. There were nine amateur radio-hams sharing the wave-bands in Westerton and vicinity, and no office-clerk's housewife ever had any difficulty getting a recalcitrant dishwasher or electric iron fixed.

It was here, in his private workshop, that he had developed his "hunch" to startling reality. Working in his spare time, figuring out its mathematical components, then working those components into theoretical diagrams, then designing and building the machine to fit the diagrams—and it worked!

Also, it was fantastic. It had been a little too fantastic for him to mention it to any of the others at the labs. His fellow-engineers—some of whom were considerably older than he was—were a little too staid for that. They, too, were products of the age; their entire efforts and, indeed, most of their interests were tied up in the one, basic problem of making better electronic devices for better weapons for the Armed Forces. They couldn't be blamed for that, the world situation being what it was, but it did make them somewhat hide-bound.

The idea of controlling the pull of gravity was a little too fanciful for them, Morrow feared—or if they had any interest in the idea at all, it would be in the possible uses of it as a military weapon. That was the way to get ahead as a scientist, these days!

Morrow shuddered involuntarily.

*There* was the thing he actually feared!

He drew it into his thoughts, slowly, and analysed it. Item: he had discovered a means of controlling gravity. Item: he had developed a mechanism which worked on that principle. Addenda: the mechanism could lift a human being, quite possibly as much mass as a heavy tank, and it might even open the way to interplanetary travel.

Quite obviously, it had terrific potentialities as a weapon of war.

And it was his patriotic duty, as a citizen of the United States, to turn his discovery over to the authorities.

Well, suppose he did? It would come as an even greater shock than was the development of the atomic bomb—of that, he was sure. It would become a top-secret project. Gradually, each individual unit of the entire Armed Forces would be made airborne. The Infantry would take to the skies, supported by airborne artillery and tanks; the Air Force and Navy would combine to send giant battleships gliding through the stratosphere, unhindered by any shore-line and capable of both artillery fire and aerial bombardment.

How could anything as big as that be kept secret? The answer was, it couldn't. Such a program would hardly have begun when some Russian agent handed the entire secret over to his bosses in the Kremlin. Then Russia would launch the same sort of program.

And world tension was already terrific. Mankind was already teetering on the brink of atomic war. What psychological effect would this new threat have on them? What insults would the diplomats think of, then? What charges and counter-charges would hurl between them? What final "incident" would spark the entire civilization into a raging holocaust?

Or would the officials in Washington realize that outcome? Would they order his discovery destroyed, forgotten, and himself assigned to some well-guarded hunting lodge in the Canadian Northwest where he could be kept in comfortable isolation, with no one around to pluck the dreadful secret from his mind?

The present balance of power had at least some promise of averting an atomic war. His discovery would destroy that balance of power, and do it suddenly, frighteningly. Someone might get just scared enough to start shooting. After that, there'd be no turning back. There would be atomic war.

They probably wouldn't want their balance of power destroyed. At least, not *that* way.

Well, then, why shouldn't he save them—and himself—a lot of trouble and simply destroy the thing himself? Forget about it, forget he'd ever thought of it?

That wasn't so good, either.

Personally, he was deeply anxious to begin the tests on the mechanism. There was so little he knew, actually, about what its limitations were, how they could be surmounted—

But there was more to it than that.

The mechanism did work, and it would lift considerable weight. Therefore, it would certainly have its uses.

Air travel could be made perfectly safe. That fact prompted a vision into his mind of everybody flying around in little, teardrop plexiglass shells, landing on their roofs—and living in homes scattered over a peaceful countryside. Cities could be smaller, devoted exclusively to office-buildings and industrial plants, and would suffer less congestion.

Also, people would become accustomed to travelling greater distances. A thousand miles might be a comfortable afternoon's ride. This, in turn, would mean greater travelling and exchange between various nations.

Then, there was the fact that commercial shipping would be revolutionized. Transporting air cargoes would be cheap and dependable, even for the heaviest kinds of freight. Thus, factories could be built near their power or raw material sources. They wouldn't have to be built near large railroad centers or harbors; commercial shipping would no longer be a problem. And thus, industrial areas could spread out, become less congested, have better surroundings for employee-morale and pay less property taxes.

Also, they would be able to ship their products to more distant markets. International trade would increase tremendously. The world-wide competition would shatter unfair national cartels—that would take time, and many governments would fight it, but eventually they'd have to accept it or intensive smuggling would undermine their economy. In times of economic stress, black markets were often a blessing to backward, underdeveloped areas.

The whole result of it would be that the entire world would be bound together far more closely. Economic ties would be predominantly international. The increased flow of travellers between nations would gradually break down prejudices and differences of custom and misunderstanding.

And that would create a far stronger basis for tomorrow's world government. As civilization stood, it needed a world government desperately. Either that, or atomic energy would destroy it. Either world government or war.

*So there it is!* Morrow concluded.

He had a mechanism for controlling the pull of gravity.

Either that mechanism was destroyed and forgotten, or the world's present balance of power would be destroyed and humanity plunged into atomic war.

But if the mechanism was destroyed, humanity wouldn't have it for the future development of world government and civilization. And they needed it. The present automobiles, trains, and aircraft were all very streamlined and marvelous when compared to the horse and buggy, but they were still too limited, too cumbersome and too costly. There had to be something for the average man, earning the average salary, that would haul him—and extend his interests—to the far corners of the world.

The mechanism would do that.

Mankind would need it to develop a sound, productive future.

But if it wasn't destroyed, there would be atomic war. There wouldn't be any future!

It was after midnight when he rose from his chair, pulled on a pair of slacks and a sweater, and left the house. He locked the front door and walked around to the garage. Swinging the door back, he felt his way into the darkness, touched the familiar surfaces of his little motor-bike, and rolled it out to the drive-way. Mounting, he kicked the starter, and the little one-cylinder, 15 horsepower engine exploded into a throaty chatter.

He rode down the dark, tree-lined streets, the cool air whipping over his body. Swinging into Railroad Avenue, he pulled over to the curb and stopped before the lighted windows of the telegraph office. He strode in, scribbled off a telegram, and paid for it.

The office girl, counting the words, stopped and frowned. She shoved it back across the counter to him. "Does that make sense?" she asked dubiously.

Morrow glanced over it again and smiled. It read:

**WESTERTON, NEW JERSEY**

**August 6, 1960**

**D. P. SMITH**
**ACME CROP DUSTERS INC.**
**DENVER, COLORADO**

**SCRAMBLE WESTERTON. WIRE E-T-A. MAY DAY.**

**BILL MORROW**

He shoved it back to her. "It makes sense, all right. And I'm expecting a quick reply, so I'll be waiting across the street in Switzer's Cafe."

"It may take some time—"

"That reply will come as quickly as you people can handle it," Morrow retorted. "A crop-dusting pilot is accustomed to getting telegrams in the middle of the night—and answering them, before some other outfit can grab the job being offered!"

The girl shrugged her thin shoulders. "All right, then. You'll be over at Switzer's—"

"Right."

She scribbled a note on a memo pad. Morrow turned and strode out.

A feeling of elation tingled through him as he crossed the street. Calling on D.P. Smith had been a natural reaction, once the plan had begun forming in his mind. If he'd ever wanted anyone murdered, Smitty was the one man he could trust!

But there was a more immediate cause for elation. It was after midnight, and Gwyn went on shift at Switzer's Cafe at midnight. She'd been on the dawn patrol for the past week, and the only time he'd seen her was when he dropped in for a quick breakfast coffee every morning.

Gwyn Davidson was the only daughter of old Pat Davidson, the plant superintendent at Western Electronics. Bill had worked under Pat as a production engineer; he'd met Gwyn two months earlier when she returned home from college. Gwyn's mother had died the year before from cancer, after a lifetime of suffering and hospital bills. Old Pat was still paying off those bills, and Gwyn had been working her own way through school. Now, she was a waitress with an M.A. degree, helping out with the expenses at home.

He saw her through the front window, leaning on the counter in the deserted cafe, reading the comics in a newspaper. She was a small, curvaceous girl in a blue waitress' uniform carefully chosen to fit to her best advantage. Soft, dark hair tumbled back from a tanned, healthy face that sported only a trace of lipstick.

Her wide, steady gaze flicked up as he strode in, then she smiled warmly. "Hi, Bill. What're you doing up at this ungodly hour?" Pretty, firm-fleshed, and bouncy.

*Even though her feet are killing her!* Bill thought. "Hello, Gwyn," he said. "I came down to send a telegram. Pour me some coffee, huh?" He straddled a stool before her.

"I'll give you what we serve as coffee," she answered brightly, "but you'll have to pay for it!"

"Fair warning. How's tricks?"

"Haven't seen her lately. What's with this telegram all of a sudden?" She grabbed cup and saucer, turned, and drew a cupful from the chrome coffee-maker.

"Invitation to an old friend," Bill replied half-truthfully. "All of a sudden, I'm lonesome."

She swung back and slid the coffee before him. Her eyes were teasing. "Wouldn't a wife do just as well?"

"A good question," he quipped back. "Come sit down and have coffee with me, and we'll talk it over!"

"*What?*" She grinned brightly, wide-eyed. "Don't go 'way, now!" She whirled, grabbed a cup and saucer, and filled it. "I'll be right there!" She moved briskly around the end of the counter and perched herself on the stool beside him. "Now! Tell me more!" She began ladling spoon-fulls of sugar into her coffee.

It was a good comedy act, done with a natural flair for perfect timing. Morrow leaned weakly on the counter, laughing silently.

Gwyn gave him a glare of feigned contempt. "Oh! Just another fast-talker, huh? I might have known!" She stirred her coffee furiously. "You engineers are all alike. If father warned me once—"

"Don't overdo it, honey," he cautioned her, lightly. "You know perfectly well I've enjoyed those long goodnight kisses when I've walked you home."

She sobered reflectively. "All right, Bill. But just what was this mid-morning telegram about—or don't you want to tell me?"

It was a casually-spoken question, and the circumstances made it a perfectly logical one. As a research engineer, Morrow worked on a number of things which had top-secret classification, and Gwyn knew he did.

*And I'd better classify this, too!* Morrow thought slyly.

"Afraid I can't," he answered her, calmly.

She nodded and sipped her coffee in silence. Finally, she asked, "Will you be glad when I'm back on a day-shift?"

Morrow took his turn sipping coffee and took his time forming an answer. "I want to take you swimming out at the Lakeshore Lodge, again," he said. "I still dream about the way you rolled up your two-piece suit so it was a Bikini model—"

"Uh huh," she interrupted. Her tone was hardly enthusiastic. "If we do, you'd better not try making the passes at me you did the last time!"

"You expect me to resist the temptation of all that beautiful skin?" he retorted, grinning down at her.

She gave a pert shake of her head. "When I give in to a man, he'll be my husband," she said firmly. "And he'll be my husband because he loves me—not because he drools over my body!"

"Ummm," Morrow ummed, doubtfully. He decided it would be best to change the subject. "Read the latest *Universe*?"

"Uh huh! What'd you think of Sturgeon's story?" She was at once bright, smiling, interested. "Wasn't it wonderful? I mean, the way he so perfectly defined an alien being's intelligence—"

That was science-fiction. Gwyn read the science-fiction magazines avidly, from cover to cover. Morrow read a few, along with his other reading—the *Post*, *Harper's*, the *Digest*, and half a dozen technical journals—and he'd even written and sold a science-fiction story once. Nineteen editors rejected it, but the twentieth bought it after having him revise it three times.

But that one mutual interest had gone a long way in winning his esteem in Gwyn's mind, slight though it was. And she was cute as a bug, the sort of female who set a man's blood a-tingle.

So they talked science-fiction. Alien creatures that inhabited other planets, trips across space and out to the other stars, travels through time and into other dimensions, civilizations which spread clear across the galaxy....

It was over an hour before a young messenger boy came in with the expected telegram. Morrow tipped the boy, excused himself to Gwyn, and ripped open the envelope.

The message read:

**DENVER, COLORADO**

**AUGUST 6 1960**

**BILL MORROW**
**WESTERTON, NEW JERSEY**

**ROGER, WILCO. E-T-A NEWARK AIRPORT 3:10 A.M.**
**SUNDAY AUG. 8TH.**
**WHERE IN HELL IS WESTERTON?**

**D.P. SMITH**

Grinning, Morrow folded the yellow sheet and stuffed it into his pocket.

"Everything okay?" Gwyn asked, forcing all concern from her voice.

"Everything is okay," Morrow affirmed quietly. "How much do I owe you?"

"Four coffees? Forty-five cents."

He laid the change on the counter, then stooped and kissed her cheek lightly. "I gotta go home and get some sleep," he murmured.

She smiled, a little wistfully. "Thanks for coming."

He went out into the cool darkness, then hurried down to the bar on the corner and went in to use the men's room. Then he came out, crossed the street, and climbed aboard his little motor-bike.

Thoughts drifted lazily through his mind as he chugged contentedly homeward....

Thoughts—and memories. They were cruising along peacefully at 40,000 feet. Morrow felt as if he were molded into the snug rear cockpit, an integral part of the tons of sleek, deadly metal that was the old F-94 jet-fighter. But he'd experienced that feeling so often it no longer mattered, then.

Before him was the familiar maze of instrument dials and signal lights and switches crammed around a glowing, green-blotched radar scope. Around him was the clear, transparent canopy, with the round crash-helmet of Smitty's head poking up from the front cockpit ahead of him. Below, off the edge of the razor-thin wing, was the criss-crossed gray surface of the Arctic ice-pack. The sky was an intense blue-black sprinkled with the hard, bright sparks of stars.

There were faint, rhythmic sounds around him. Familiar sounds. The warm, dry air blowing through his flight suit, circulating over his body. The air pushing into his face-mask. The rolling motion of the seat-cushions, massaging his backside with mechanical dispassion.

Then the flat, metallic voice in his earphones. *"Forty-three degrees left. Contact in five minutes!"*

*"Roger!"* Smitty's voice answered.

The ship tilted gently. Centrifugal force pressed Morrow against his seat. The world turned slowly beneath them. Forty-three degrees.

Two minutes later, a bright spark appeared on his radar scope. *"Air spotted!"* he spoke into his mike. *"Two degrees right!"*

*"Over to you!"* the metallic voice from ground radar answered. And the jet shifted slightly. Two degrees.

*"Contact in two minutes,"* Morrow chanted. *"One-thirty ... One ... Thirty—"*

*"Contact!"* Smitty's voice cracked.

The F-94 whipped over into a turn. The force of two gravities shoved Morrow down in his seat.

For a brief moment—a breathless, eternal moment, all of two seconds—another F-94 exactly like theirs appeared directly before them. Long enough for red lights to glow and camera guns to record a direct hit. The practice mission was completed—almost.

Then Smitty snap-rolled the ship, missing the other ship almost by inches. The g's piled up, cramming Morrow down in his seat, pulling at his facial muscles. Then his vision cleared and he straightened up, bruised and somewhat battered.

It was the old bomber-interceptor game. That other F-94 could have been an enemy bomber, plowing toward American cities with a load of atomic death—

Smitty turned his head and looked back. His eyes crinkled into a smile under the green glaze of his goggles.

Smitty. Captain Daniel Purcell Smith, then—or "D.P." Smith, which were also the initials for "Displaced Person." A cool, thoughtful, and smart jet-fighter pilot in those days, and a darned good guy. They had taken

Seattle apart at the seams on their one furlough, preferring the devilment of their own companionship to going home to Mom's apple pie.

Morrow's telegram had made sense, all right. The words *scramble* and *May Day* were fighter-lingo; *scramble* meant *let's go! we've a fight on our hands,* and *May Day* meant *I'm in trouble!*

He was in trouble, certainly. The mechanism he'd developed was, in itself, plenty of trouble.

And it was a special kind of trouble—the kind in which the only person he could dare trust had to be someone like Smitty. The Air Force camaraderie which existed between them had never quite faded out. Even after they'd been mustered back into civilian status, after Morrow had signed a government engineering contract and Smitty had gone on to commercial flying, they had kept in touch with each other. Diverging interests hadn't pulled them apart; the old school ties, the old trustworthiness was still there. An odd letter every few months or so, a postcard at Christmas....

He was fortunate to know a man like Smitty, Morrow knew. He couldn't have carried out his plan alone.

He reached home, stored his motor-bike in the garage, and walked into the living room. He snapped on the light and stood there for a moment, gazing across the room at his littered writing desk. If he were going to carry out his plan, there was one thing he'd have to do himself. People weren't going to like it. Good engineers were scarce.

He walked across the room, sat down at the desk, and crammed a sheet of Western Electronics stationery into his portable typewriter. He paused, lighted a cigarette, and then grimly proceeded to write his letter of resignation.

"It's a mechanism that floats in the Earth's field of gravity," Morrow began—

They were seated in a secluded booth in the modernistic restaurant at the Newark Airport. Through the wall-length observation window, they could look down on the airfield; a giant stratoliner was rolling up before the building, the bright spot-lights glistening off the silvery arcs of its six big turbo-props. White-uniformed linemen were pushing the steps up to the side of its fat hull as the door slid open and a pert stewardess poked her head out. Beyond the gleaming sky-monster, in the pitch darkness of early morning, the runway lights twinkled in rows and patterns of red, yellow, blue and green sparks.

Morrow spoke quietly and succinctly, pausing only for a sip of coffee or a pull on his cigarette, and gave a concise briefing of his discovery and its implications. The dishes of an early breakfast had been cleared away, so no waitress bothered them and the few other patrons in the restaurant were out of ear-shot.

Across from him, ex-Captain D.P. Smith sprawled laconically on the cushioned seat, listening. The expression on his lean, brown face was thoughtful, intent. He sipped his coffee and flicked the ashes of his cigarette into the saucer.

He was a small, slender man dressed in a conservative, pin-stripe business suit. There was nothing dare-devil about his attitude, nor were his movements deft or quick. He was slow, cautious; his attitude was a reserved calmness.

It was immediately noticeable. His carefully groomed black hair and his small, black mustache gave his features a mischievous look. There was something satanic about his small stature, his long hands, and his lean, handsome appearance. One would expect a bright, hand-painted tie and a roving, speculative eye. His utter calmness and reserve seemed incongruous.

Only the faint, white scar along his jawline might have indicated a devil-may-care experience. Morrow had mentioned it, remembering that Smitty had written about the crash last year—he was making a pass over a field, spreading bug-killer spray over a farmer's potato crop, when a sudden down-draft caught his plane and he couldn't pull up in time to avoid the neighboring orchard. He'd crashed through the apple trees, snapping them like kindling. The plane was completely demolished.

*When I woke up,* he'd written, *they had me spread out on a silver tray with an apple in my mouth!*

Crop-dusting was a hard, dangerous job. The pilots did most of their flying before dawn or in early afternoon, when the air was calm; but they had to fly at other times, too, to make enough to meet expenses. They'd take off in small, worn-out planes, loaded beyond safety flight limits with bug-killer, and fly to some farmer's fields. Then they'd make passes back and forth over the fields, flying below-treetop, leap-frogging barbed-wire fences, zooming under telephone lines, and dodging trees and farm buildings, their eyes stinging as the spray billowed back into the cockpit.

The pay they received was small, mostly because there were so many skilled pilots looking for work and so few civilian flying jobs. Smitty could easily reenlist in the Air Force, of course, but they wouldn't give him a flying job; at thirty-two, he was too old for military flying. They took the eighteen-year-olds for that. And Smitty wouldn't reenlist to sit behind a desk.

So he dusted crops. It was no job for a dare-devil, either. A pilot had to know his limitations, the limitations of his plane, and what he was doing every second.

"—And that's the situation," Morrow concluded. "If the mechanism isn't destroyed, it'll plunge the world into atomic war. If it is destroyed, it'll be lost to mankind for the next several hundred years—until somebody else stumbles across it."

"In short," Smitty resumed, "if we got it now, we have atomic war. If we don't have it for the next few centuries, we *will* have atomic war."

"I'm afraid so," Morrow affirmed. "Unless they manage to develop a world civilization and government without it."

Smitty shook his head. "They need something like this gravity machine to pull people closer together, to get them to know more about one another. Otherwise, any world government scheme is likely to be a fizz—unless it's established by force!"

"That'd amount to world dictatorship."

Smitty shrugged. "All right, so we've got this thing. If we keep it, we get atomic war. If we don't, maybe our grand-children get atomic war. That it?"

Morrow nodded.

"So you must have some plan up your sleeve!" Smitty grinned at him, shrewdly. "You wouldn't drag me all the way up here just to listen to a hard-luck story."

Morrow's eyes narrowed. "Smitty, the only reason this would cause an atomic war now is because the world situation is so tense—"

"True!"

"—But the world situation isn't always going to be this way! Sooner or later, something will happen to change it. Something's bound to change it! This is a modern, fast-moving world—things happen fast!"

"So?" Smitty raised his brows, querulously.

"Well, it's bound to change within our lifetime! And when it does, we may have an opportunity to reveal this discovery. All we have to do is wait, keep it secret, test it and develop it, and turn it loose when the time is ripe!"

"Un-huh," Smitty grunted. "And who's going to pay for it?"

"I've got seven thousand in the bank—"

"And I've got three!" Smitty frowned scornfully. "How far do you think we'd get on ten thousand bucks, chum?"

"As far as we'll need to get," Morrow retorted. "We aren't trying to finance a mass-production scheme, remember. This is strictly experimental work."

"What would the retail cost amount to on that mechanism you built?" he asked dubiously.

Morrow scratched his jaw, reflectively. "Retail cost it'd run to around three hundred dollars."

"So we make a bunch of those mechanisms. Now, what do we test 'em for?"

"For their use as a means of air transportation," Morrow answered. "Primarily, that is—there are probably a good many other possibilities."

"So how do we test 'em?" Smitty persisted. "How do you test any flight mechanism? You take it up in a plane, turn it on, and see how it works! So for thorough tests, including high-altitude performance, we'll need a plane with a pressurized cabin, big enough to hold our test equipment and the mechanisms. At the present market rates, you won't buy a plane like that for much less than fifteen thousand dollars!"

Morrow was shaking his head, patiently. "We can't do it that way," he said. "But we can afford the cheap plastic materials they're using in small private planes, now, and build a ship especially for the mechanisms. Then we can test it for low-altitude performance and, if it works, gradually extend our tests on up to eight or ten thousand feet—"

"And if the mechanisms fail, we crash! That'd be sheer suicide—"

"Not necessarily. If they work at low altitude, they'll be dependable in saving us from a crash. And we can install a main and auxiliary system of mechanisms, so if one fails we can cut in another."

Smitty paused, thinking it over. He gave a slow, grudging nod. "It might work, at that. It just might. But you realize what sort of predicament this will put us in, don't you?"

"Such as what?" Morrow prompted cautiously.

"Such as supposing somebody finds out about it," Smitty replied. "Most people have a pretty strong feeling about patriotism these days. We have something that qualifies as a good secret weapon. They aren't going to like the way we neglect to inform the government about it."

"Uh huh. Men have been lynched for less," Morrow agreed. "We'll just have to see to it that nobody does find out about it. We can start out small, in almost any place that's relatively isolated—a deserted farm-house would do, I suppose—and build our ship. Then we'd have to make our flights at night, until we're fairly sure of the ship. After that, we could set out to find a permanent base—one hidden off somewhere in the desert or mountains, where nobody will notice us. Then we'll fly our equipment out there and set up shop."

"What about power? If we set up near a power line, there'll be the company linemen coming around."

"I think a gas-engine generator will suffice," Morrow refuted. "We can haul gas to our deserted farm-house by car, then fly it out to our shop at night."

"What if somebody asks questions when we buy or lease this land, 'way off in the middle of nowhere?"

Morrow grinned. "If it's 'off in the middle of nowhere,' why should we buy it? Nobody'll know we're there!" He finished the last of his coffee and shoved his cup aside. "You've been flying over the Southwest for quite some time, Smitty. I'm hoping you can find the sort of isolated spot we'll need."

"There are places in that desert country where no white man's ever walked," Smitty confirmed. "They're still finding old Indian ruins nobody knew existed. But you know we could get arrested for all this, don't you?"

"Umm," Morrow ummed. "Building an experimental aircraft without authorization *is* unlawful, isn't it!"

"It's a federal offense!" Smitty exclaimed tersely. "Also, flying without a license is a federal offense—and you don't have one. And using government land without permission is a federal offense. And you'll have to quit your job with Western Electronics, won't you? What about your government contract?"

"I've given them two-weeks' notice," Morrow explained. "I'm allowed that. Of course, engineers are scarce—so scarce that by quitting my job here for no good reason, I'm getting myself blackballed out of every other company in the industry. None of 'em will hire me after that."

Smitty frowned concernedly. "Did you have to do it that way? I mean— suppose you just disappeared?"

Morrow shook his head. "There'd be federal investigators swarming around here three-deep!" he said. "I repeat, chum—engineers are scarce!

And they don't like strange things happening to engineers who've been working on top-secret material. They catch more enemy agents that way."

"You sure they won't investigate you for quitting?" Smitty's gaze was thoughtful.

"I don't think so. In the next two weeks, I think I can convince them that I've simply turned out to be a stinker." Morrow grinned sourly. "They'll be glad to get rid of me, then."

"So you'll be ready to leave in two weeks." Smitty's tone was non-committal. "Then I'd better hop the next plane out this morning and start hunting up our base of operations."

"Don't you want to come out to Westerton and see the mechanism?"

"Uh-uh! Less we do to arouse suspicion, the better. I'll wire you, of course, when I find something. Have you got a gun?"

"Gun?" Morrow started. "No. Why should I?"

"Good." Smitty grinned lazily. "Don't carry one. They're too damned dangerous."

"I agree," Morrow said quietly. "It hadn't even occurred to me."

The train rattled and squealed through the hot summer afternoon, dust and foul-smelling smoke drifting back through the open coach windows. Morrow huddled in the corner of his seat and stared miserably out at the moving landscape.

*Have you got a gun?* The words echoed through his mind. Of course he didn't have a gun. He had never thought about it. Why should he need a gun?

But the answer was obvious. The secret of the gravity-control mechanism was precious.

Certain individuals, should they learn about it, would stop at nothing to get it. Including murder.

And if the government learned about it, they'd dump him into prison and throw the key away!

Thus, anyone who happened to find out about it would do one of two things—try to steal it or inform the authorities about it. Either one would mean catastrophe.

And there was only one sure way to keep anyone's mouth shut. Kill them!

Morrow knew he couldn't do that—he didn't have that sort of mentality. Nor could he stand by and let anyone else do it, not even Smitty.

But that was what Smitty had meant: he wouldn't stand by and let it happen, either.

Besides, any murder would bring on an investigation. They couldn't hide from that. So it boiled down to the simple fact that if anyone found out what they were doing, they'd be finished. Dead men don't talk, but they get a lot of other people curious.

Somehow, they had to keep it secret. They couldn't afford to let anyone find out about it.

And that could be disastrous. There had to be some alternative choice, in case anything like that should happen. There had to be an out. Without one, they'd be trapped.

They had to admit that some day, somehow, it *would* happen. Someone *would* find them out. And they had to be prepared to handle it. It would have to be handled in some way that didn't involve murder.

What other way *was* there?

There had to be some other way. *Had* to. Morrow chewed down his fingernails as the train lurched and rattled onward....

They pulled into Westerton with a hissing roar of steam and jolted to a stop beside the station. Morrow climbed down from the coach, wearily, and strode through the station to the street. It was late afternoon, but it was still hot. He pulled off his tie, stuffed it into his coat pocket, and unfastened his collar. Then he pulled off his coat, threw it over his shoulder, and rolled up his sleeves. That was better. Now for a bite to eat.

He strolled down the shady side of Railroad Avenue toward Switzer's Cafe.

*Beyond the law!* his footsteps rang on the sidewalk. *Beyond the law, beyond the law—*

Suppose someone did find them out? They could ask no one to protect their interests. There'd be no help from the authorities. They'd have to protect themselves—against anyone and everyone! How could they do that without guns, without the possibility of killing someone? They couldn't accept defeat that easily. The secret was too important to the future of mankind!

But what could they *do*?

*Beyond the law! Beyond the law—*

"Bill! Hey, wait up!"

Morrow stopped as if someone had jerked him back on a string. He whirled toward the sound of the voice before his mind had recognized it.

Gwyn came trotting down the street toward him, swinging a tennis racket in her hand. She was dressed in a white, short-skirted tennis suit. She stopped beside him, breathlessly, and put her arm through his. "Where you going?"

"Switzer's," he said. "Join me in a sandwich?"

"Okay." They strolled onward. Her skirt rippled over her smooth thighs, accentuating her tanned, slender legs. "I go on the four o'clock shift tomorrow. Want to come down at midnight and walk me home?"

"At *midnight*?" he taunted.

"Sure! It's the witching hour!" She wrinkled her nose up at him, teasingly. "What're you all dressed up in your suit for? Going somewhere?"

"Had to go to Newark today," he said. "To meet someone."

"Oh! Don't they even let you alone on Sundays?"

"Sometimes, honey." He grinned. "When are you going swimming with me again?"

"Well, if you want to *swim*—" She broke off and gazed up at him with mocking cynicism. Suddenly, her gaze went past him and she tugged at his arm. "Oh! Wait a second."

She guided him into the little newsstand and left him by the cigar counter, going on over to the magazine racks. Morrow stood back and admired her firm, shapely posterior.

"Ah!" she exclaimed, pulling out a magazine. She fished some change from the little purse on her belt and passed it to the newsstand operator. "Okay, let's go."

"What've you got there?" Morrow asked.

"You can see it after I have," she retorted. "Why don't you buy one yourself, for a change?"

She flipped through the magazine's pages as they walked along. Morrow took her elbow, guided her around a telephone pole, and maintained a discreet silence.

As they seated themselves in a booth, Gwyn closed the magazine and slid it across to him. Smiling, Morrow glanced down at it—then stiffened, staring at the cover illustration.

It was no more than a typical science-fiction cover. The setting was a typical street scene at night—some dark side-street in the metropolitan section of some city like New York. In the foreground stood a young man....

But from there on, it was nothing ordinary. The young man was slumped back against the wall of a building as if he were trying to mold himself right into it. The expression on his face was one of mixed surprise, incredulity, and fear. It showed plainly that he knew no one else would believe him if he told what he was seeing; and furthermore, he didn't believe it himself.

In the background, farther up the street, a group of people were emerging from a doorway. A beautiful girl was in the lead, and behind her came creatures that looked like men with blue skins, except that they had tentacles instead of arms. The light of a street lamp revealed the skin-tight garments they were wearing, and the octopus-armed men had transparent helmets over their hairless heads. The girl wore a helmet that was thrown back.

And before them was a tall, gleaming rocket ship, standing on its tail-fins in the middle of the street!

And the young onlooker *didn't believe it*!

"She *is* pretty, isn't she!" Gwyn's acid tones cut through his thoughts.

Morrow noticed, then, that the cover-girl's costume was not only skin-tight, but there wasn't much of it. He grinned wordlessly, then thumbed through the rest of the magazine. Its pages hardly registered on his mind. He was beginning to form an idea....

By the end of the following week, Morrow had convinced everyone at the labs that he was a heel. But that wasn't all. He also *felt* like a heel.

It began the first day, with Borgesdorf. Alec Borgesdorf was chief of the Research Division. He sent word for Morrow to drop into his office. When Morrow walked in, he saw his letter of resignation on the desk. Borgesdorf was grinning and frowning at the same time.

"What the hell *is* this, Bill?" he asked good-naturedly.

"What's it look like, 'Greetings from the President?'" Morrow retorted.

Borgesdorf's grin faltered. His frown turned to amazement. "Well holy cow, Bill!" he exclaimed. "What's the trouble? Why're you quitting?"

"I'm quitting this whole blasted mess!" Morrow said flatly. "Does that answer your question?"

"Wh—well, yes, if you say so. But—you know what this means, Bill! *Why?*"

Morrow looked at him, coldly. "Suppose you mind your own business?"

Borgesdorf tensed behind his desk. The friendliness faded slowly from his gaze. "All right," he said abruptly. "But if there's anything wrong around here, I think you should tell me about it."

"Don't worry about it," Morrow sneered. "I'm quitting and that's that. Keep your dirty nose out of it."

Borgesdorf's big, fleshy face reddened slightly, but that was all. He didn't say anything for a few minutes. Then he gave a barely perceptible nod. "Very well, Morrow. That's all."

"Sure." Morrow wheeled and stalked out.

Two days later, it was little Petersen. Petersen was a wizened, little guy nearly sixty years old; he'd been playing around with radio when it was a crystal and the cat's whiskers. He had consternation written all over his seamed face as he came shuffling up to Morrow.

Morrow could almost hear the discussing that had gone on between him and Borgesdorf—Petersen frowning worriedly as the chief said, *I couldn't get a thing out of him, Pete. Can't understand it at all. See what you can get out of him, will you?*

So here was little Pete.

"Hear you're quittin' us, Bill," he drawled nasally.

"What about it?" Morrow retorted, cursing himself mentally. Pete was a nice, old guy—everybody in the labs liked him. Morrow liked him, too ... but this was different.

"Nobody's done anything against you, have they?" Pete complained. "You're throwing away a whole lot, son. It won't be gotten back easy." His shrewd, little eyes watched Morrow, pensively. "The country needs young fellas like you now, Bill—"

Morrow forced the sneer across his face again. "That's just too damn' bad," he said evenly.

Pete's eyes narrowed. "You're talkin' like a commie—"

Morrow lashed out. The back of his hand smacked across the little man's mouth. "Beat it," he said huskily. "Beat it, you damned little shrimp."

Pete stared at him for a moment, then turned slowly and walked away.

Instantly, Mart Sumter came stalking across the lab. Sumter was big, broad-shouldered, with muscles bulging against his stained smock. He stopped in front of Morrow, his fists clenched.

"If I ever see you do that again," he said softly, "I'll give you the worst beating you've ever had in your life!"

Morrow returned his angry glare, then whirled and went back to his work.

"You heard me, didn't you?" Sumter's breath whispered on his neck.

"I heard you," Morrow rasped.

"Don't forget it." Then Sumter strode away.

Morrow grinned shakily. He was certainly getting what he deserved!

At home, an idea was rapidly taking on form and dimension in his mind. He set up his drafting board, collected his inks, and worked doggedly through the night, etching out diagrams that showed—theoretically, at least—how his idea would work.

At midnight, he would show up at Switzer's Cafe to walk Gwyn home.

The nights were cool and pleasant, with deep shadows along the tree-lined streets and the street lights filtering through the treetops, dappling the silent fronts of the houses. They strolled along, slowly, their arms around each other, Gwyn's body pressed close to his.

"I like a small town," Gwyn murmured softly, one night. "'Specially at night—so peaceful, so cozy."

"I like the dark," Morrow said.

"Why?"

"I don't know. It changes things. It's a different world."

She looked up at him, wonderingly. "I think of a small town. You think of a different world. Why is that, Bill?"

"You're tired, maybe." He grinned down at her. "You've been on your feet eight hours."

"That makes me think of a small town?"

"Contentment," he said. "Small towns are contented."

"And a different world—that's exciting, isn't it?"

"Sometimes it's dangerous."

"I see." She was quiet for a while. Then, "I never asked you where you were from, did I?"

"No."

"Small town? Or city." The latter held conviction.

He chuckled. "You're not even warm! Casa Verde, Arizona. A cluster of shacks in the middle of a desert, with sand-stone cliffs rising like mountains of the moon everywhere you looked, and black buzzards circling in a hot, brassy sky—"

She shuddered. "It sounds terrible."

"—And beautiful." He murmured it, gently. "We left when I was six years old. No schools there."

"Then—you're *from* a different world, is that it?"

"You might say that."

"Strange. We're two utterly different people, aren't we, Bill?" She was gazing up at him, studying his features, watching the dappled light and shadows play over them.

Morrow sensed that he was on perilous ground. He said nothing.

"You aren't happy here, are you, Bill?" she spoke almost in a whisper. "You never will be!"

"Most men I've met are—searching for something," he replied hesitantly.

"But they don't devote all their time to it," she protested. "They at least manage to live fairly normal lives and raise families—"

"Do two utterly different people—" He broke off, leaving the question unspoken. *But we're down to brass tacks, now,* he mused. *We just don't feel the same way about things!*

Why was that?

"Look," he said, almost gruffly. "I think of a different world—let's stick to that point, for now. You think of a small town. But then, why do you read science-fiction?"

She frowned in puzzlement. "What do you mean?"

"Well, isn't that a 'different world?'"

"But it's *fiction*. I don't think of reality—"

He smiled gravely. "You don't think there are horrible monsters lurking in the corners, or little people in the wallpaper, or strange eyes floating around watching you—"

"If I did, I'd be in a booby-hatch!"

"Or you might be a research engineer!" He chuckled softly. "Ignorance is bliss, Gwyn. And how much do you know about reality? What do you know of the mysteries within the atom, or the strange way the Universe seems to be expanding as if it had exploded and the stars, including our sun, were still hurtling outward from the blast?"

"Sounds like a good way to go crazy!" She looked up, intently. "I've never heard you talk like this."

"Well, you aren't quite so ignorant," he amended teasingly. "You realize inwardly that your 'small town' isn't quite so contented as it seems—that beneath the surface, there's unrest. So maybe you read science-fiction because it deals with spectacular forms of unrest—men risking their lives in space travel or on other planets, changes and developments that cause revolutions or wars—and you find solace in that. The little unrest in your 'small town' no longer seems so bothersome."

"Mr. Morrow," she spoke icily. "I do *not* enjoy having you pick my mind apart!"

*Then why must you criticize me?* he thought. But he didn't say it aloud....

Sunday afternoon, they went swimming. There was a secluded strip of beach where Morrow spread a blanket out on the sand, and after they had swum and splashed and dived to near-exhaustion, they sprawled themselves out on the blanket and let the warm sun dry their skin. Gwyn lay on her stomach and removed her halter, then rolled her trunks into a narrow band around her thighs. Morrow watched with mingled interest and affection. Gwyn scowled at him, then pretended to ignore him.

When his skin began to sting through the sun-tan oil, Morrow suggested they move into the shade of the trees. Gwyn struggled back into her halter and sat up. They dragged the blanket back into the shade and sat down again. Morrow put his arms around her, and they talked for a while.

When Gwyn came out of the bushes wearing her shorts and blouse, she grinned and wrinkled her nose at him. "This has been wonderful, Bill. I almost wish we could be like this forever!" She let him kiss her, then.

They rode back to town on his little motor-bike and had cokes and hamburgers at a lunch-stand.

The second week passed without significance. The other engineers at the labs treated him coolly, now. They'd be glad when he left. At home, his diagrams were finished. He went over them again, checking them thoroughly.

Friday, a telegram reached him at the labs.

STOCKTON, CALIFORNIA

AUGUST 20 1960

BILL MORROW
WESTERTON, NEW JERSEY

HIRED PLANE AND FLEW RECON. PERFECT SITE
LOCATED NEVADA. LEASED
ABANDONED SAWMILL IN SIERRA NEVADAS NEAR
HERE. WIRE E-T-A TO L.A.
INTERNATIONAL AIRPORT. MEET YOU THERE.

D.P. SMITH

Saturday, he spent most of the day settling his affairs and packing for the trip.

The plan had begun. Whether it worked or not, he was going through with it.

The gravity-control mechanism would not be turned loose on the world to increase the tensions and fears already much too prevalent to the point where mankind would plunge into atomic war.

But the gravity-control mechanism wouldn't be abandoned, either. They'd develop it, secretly. They'd have it ready when the world situation changed—for the better, Morrow hoped—and it could be given to the world. Then, mankind would benefit from it.

That was his whole purpose—that his discovery should benefit mankind, rather than pave the way to the destruction of civilization. Morrow considered it a purpose worthy of all the sacrifices he had to make. His job, his career—perhaps later, if the plan worked, he could regain them.

But he had to try.

The battered, worn truck came whining out of the rutted dirt road, clashed its gears, and rumbled into the wide sawmill yard. On the left, a little mountain stream laughed merrily over the rocks and then widened out, ahead, and trickled sluggishly across a brakish pond. On the right, at the foot of the tall pine trees, were the crumbling ruins of sheds and outbuildings, piles of rotted wood.

The truck halted before the main sawmill building across the yard. The mill was weather-stained and decrepit-looking, with the boards fallen off one wall and the roof sagging on one corner, but it was still standing.

Morrow and Smith climbed down from the cab of the mud-splattered truck and stood gazing around them. "Looks like she's been abandoned for quite a while," Morrow remarked noncommittally.

"She has," Smitty agreed. "But it'll serve our purposes, I think. This main building is large enough to be our hanger-workshop with a minimum of repairs. The timbers have tightened up until they're like iron, else the whole building would've collapsed long ago."

Morrow nodded. "Long as the timbers are sturdy, we can patch up the holes with canvas tarpaulin if we have to."

Smitty hooked his thumb toward the stream. "I got a lab analysis of the water—it's drinkable. But we'll have to spread some oil on that pond to kill the mosquitoes. We don't have any neighbors to worry about within ten miles of here; Yosemite National Park's due south of here about fifteen miles."

"What about fire towers?"

"Forest rangers? The nearest is over on a mountainside twenty miles away. He's not going to see anything at night unless it's a fire." Smitty grinned reflectively. "I figure this can serve as our temporary base until we get the ship built and flight-tested. Then we travel due east across some of the worst desert and mountain country you'll ever see, to the site I've picked for our permanent base. It's in a deep, crooked canyon over on the other side of the Kawich Range, in Nevada."

"Not near any atomic project area, is it?"

"Uh-uh. Not near anything else, either. It's not near any airway routes, and private pilots shun that area because there aren't any fields or meadows available for emergency forced-landings."

"Sounds good!" Morrow complimented him. "Where do we camp, here?"

"I've knocked together a small cabin back in the woods. Grab your stuff out of the truck and come on—I'll fix us some chow!"

Morrow climbed into the rear of the truck and slid his luggage back to the tailgate. Smitty took a couple of suitcases, Morrow the third and his equipment case, and they strode off on a narrow trail winding through the trees.

"Now what was it you've been working on?" Smitty asked as he led the way.

"I've been working, on?" Morrow echoed blankly, his mind filled with sensations of clear, cool mountain breeze and the smell of tall pines and the eternal silence of the woodland.

"Yeah!" Smitty prompted. "When we were having dinner, back in L.A., remember, we were talking about the event of anyone catching us at this, that we'd be finished if they did? You said you'd been working on something that would protect us from discovery."

"Oh, that!" Morrow grinned. "I merely figured out a means of camouflage."

"Camouflage?"

"It's still just in its theoretical stage, but I think it'll work. I'll show you my diagrams."

"Show me while we're eating."

The little shack nestled under the pines was cozy and weather-proof, built out of rough lumber and fitted out with hand-made furniture. The air was filled with the aroma of fried bacon, coffee, and wood smoke. They sat at the small, wooden table and ate out of tin plates, washing it down with tin cups of coffee, and Morrow spread his diagrams between them and explained his idea to Smitty.

"—So it's all designed around that propulsion unit," he said. "The gravity-control ring establishes a focus of 'false gravity' inside the tail-pipe so that air is sucked in through the scoops on the ship's hull. The air 'falls' into that focus of 'false gravity' and goes on past it to shoot out the tail-pipe at an estimated sixty-mile-an-hour gale."

"Couldn't we do just as well with a large electric fan?" Smitty asked, half-jokingly.

"This propulsion unit will cost only a fraction of the price of a large air-conditioning fan and motor," Morrow pointed out.

Smitty grinned at the diagrams. "Okay, but you've certainly sketched in a fancy-looking ship, there! Aero-dynamically, I'm afraid it wouldn't be too practical—"

"I know," Morrow admitted. "But we'll have to work that out and still keep this fancy-looking ship."

"How come?"

"Because that's the whole idea, Smitty! Think a minute. Suppose we've built our ship and are flight-testing it. So there's always the possibility that someone will see it—"

"—And call the cops!"

"Right. Normally, that would bring on an investigation and we'd be finished."

"I hope they serve good food at Leavenworth!"

"Stop interrupting, will you? Now, the idea is this: suppose whoever sees us *thinks* they're seeing a ship *from outer space*?"

Smitty's grin faded. He stared at Morrow for a moment, then picked up his cup and took a healthy swig of coffee. "I see what you mean," he said, replacing his cup carefully on the table. "They think they're seeing a rocket ship from Mars, or something like that. So they go to the cops and start yelling about it. And that's happened so often—"

"We won't have to worry about any thorough investigation," Morrow concluded, smiling. "They might check the area in which our ship was sighted—"

"Which isn't likely to be around here!"

"—But that's all. Even if it is around here, they aren't going to ask us too many questions so long as we don't have two heads, blue skin, and arms like an octopus!"

Smitty chuckled mirthfully. "You'd better keep out of sight, then!"

"Cut the quips!" Morrow growled mockingly. "I think the idea will work. We'll just have to design the ship so it looks weird enough to excite the imagination. It may have some aerodynamic faults, but it's worth the trouble."

"We can't make it *too* fancy," Smitty warned. "It's still gotta fly!"

"We don't want it too fancy—just so it *looks* like a spaceship! First thing we'll have to do, though, is check the costs of plastic construction materials for aircraft." Morrow gulped the last of his meal down with a swallow of coffee, stacked his cup, plate, and utensils, and set them aside. "We don't want to go too deep into our capital to build this ship," he said wryly. "The lease on this property has already soaked us two thousand."

"What'll the shop machinery come to?" Smitty asked pensively.

"Around a thousand, I think."

"Then I think we can build the ship for around—well, anywhere from one to three thousand dollars. At the most, that'll be just over half our capital down the drain." He frowned. "What'll the rest of it be for? Operating expenses?"

"Mostly that. There are a few other ideas I'd like to try out, though—experiments with these mechanisms. But remember that we're dedicated to this thing until the world situation changes and we can turn it loose without any risk. That may not come for years!"

"I've thought about it," Smitty retorted, grinning. "There's a deer run over near our Kawich mountain hide-out, and other game is plentiful. Our meat supply for the next hundred years costs no more than the price of a couple of hunting rifles."

Morrow shook his head. "That might be fine, Smitty. Maybe we could plant a vegetable garden, too, and live off the land. But I don't think we should subject ourselves to the life of a hermit. We've got to keep our perspective with this thing, and not get anti-social about it."

"A hermit's life would get kinda boring, anyway," Smitty conceded. "But I can always go back to crop-dusting and make a few dollars now and then. What'll you do, though? Can you get a job?"

"I know electronics!" Morrow smiled grimly. "I suppose I could open up a little radio repair shop somewhere."

"You? A radio repair shop? The first real genius this country's had for—" Smitty broke off, staring at him.

Morrow stared back, scowling. "Genius?" he echoed. "What in hell ever gave you *that* idea?"

Smitty grinned faintly as he lighted a cigarette. "Guess I'm just carried away by your two heads," he said, spewing smoke.

It was a full month's work just to purchase the shop machinery, the building materials to patch up the old sawmill, the materials for the ship's construction, and to truck it out and install it in the building. They worked from daylight 'till dark, then retired to their shack and spent most of the night going over the blue-prints for the ship. Gradually, it took shape and form on paper.

Masses of cloud were banked against the surrounding mountains, covering the sky with a solid, gray mass that shook loose a thin drizzle of rain, just enough to dampen the ground, the morning they conducted the first weight-test.

They used the gravity-control mechanism—they called it a *gravitor* by then—which Morrow had built in Westerton. The test was conducted outside, with a sling suspended under the gravitor to support a pile of sandbags, with a rope hanging from its bottom to a small hand-winch on the ground.

The gravitor rose up into the drizzle with its load, lifting three hundred and sixty-nine pounds to a height of forty feet. It floated there, the rope dangling loosely from it. There was an odd three-foot S-curve in the rope just below the sandbagged sling.

Smitty stared up at it, squinting against the misty rain. "It just floats there!" he exclaimed huskily. "On four flashlight batteries—"

"The wind's drifting it toward the trees," Morrow said in a tight voice. "Better take up the slack."

Smitty stooped and wound up the little hand-winch. Then he straightened and stared upward again. "On four batteries," he repeated in his husky murmur. "Look at that snake-twist in the rope!"

"That part of it's inside the gravitor's field," Morrow explained quietly. "As for the batteries, I think it's because the mechanism is shielded from the gravity and magnetic influence of the earth. It works entirely within its own magnetic field. Its electronic conductivity is more efficient, so we're getting far more power from those flashlight batteries."

"But *is* there that much power in a flashlight battery?"

"Don't forget those batteries are also inside the gravitor field," Morrow reminded him. "Anyway, I'm not even sure that's the answer. The scientific implications of this extend to such matters as the dimensions and volume of the Universe, and the speed of light. Maybe the Universe isn't expanding and maybe light 'particles' or 'congealed energy' or whatever they are don't slow down. Maybe they curve through a kaliedoscope of gravitational forces generated by star-clusters, and the 'expansion' is a matter of refraction in our particular sector of space—"

"Do you have these attacks often?"

Morrow looked down to find Smitty watching him with a mocking leer.

"C'mon, professor," Smitty chided him. "Let's crank this thing down and get in out of the rain."

"Ummm? Oh—all right!"

Crude wooden jigs were sawed out and nailed together. Plastic tubing was heated and curled into the jigs and, when cooled, was taken out in the precise shapes of formers, spars, and bulk-head frames. These were welded onto thick plastic rods and the rough outline of the ship began to appear. More rods were added, strengthening the framework, and the ship began to assume its final shape in a spidery basket-work of glistening, transparent plastic.

The covering was torn off a large roll of celatex film, and long strips of it were spread through the inside of the framework and cut to size. The strips were dipped in a softening bath, then stretched across the inside of the framework, pressed against it and, drying, molding to it to form a tough, rigid inner skin. Fistfuls of plastic insulating material was dipped and sponged into the openings in the framework, molding to it and to the inner skin. Then more strips of celatex were cut to size over the outside of the framework, dipped, and stretched over it to form a strong outer skin. The result was a large, sleek hull, with a shimmering basket-weave framework and frosty-white, fuzzy insulation showing through its transparent skin.

Gravitors for the lift units and the propulsion unit were built, tested, and installed. A cargo deck was built into the belly of the ship, accessible through large side doors. Power circuits and control systems were installed. The forward, control pit, and aft compartment decks and bulk-heads were welded into place. Then they let their imaginations run riot on the interior decoration, fittings, and furnishings which were easily constructed of plastic framework with celatex stretched and pressed firmly over it to form the desired curves, bulges, and flowing lines. Then they went over it with sand-paper, paint-brushes, and dark blue and mirror-chrome plastic lacquer.

The interior was, to put it mildly, luxurious and ultra-modern. Smooth, flowing instrument panels and storage lockers molded into the walls, foam-rubber chairs growing out of the decks, bunk-seats sunk into the bulk-heads, and transparent-topped tables sprouting their chrome frames from the fore and aft lounge decks. They finished it up with a small lavatory and an electric hot-plate in the bulk-head cubicles just off the forward lounge.

Finally, transparent plexiglass was fitted into the long port-hole slots along the hull, and a large plexiglass dome was mounted over the control pit above the smoothly tapered nose. Then they papered the plexiglass and manned a spray-gun, giving the entire outer skin a thorough coat of shimmering black lacquer.

The complete construction took all of six weeks working from dawn to well after sunset. When it was finished, Smitty took the truck and went into Stockton to purchase the three automobile batteries which would be used to power the ship.

That night, Morrow sat at his drafting table scrawling rough diagrams and pencilling in mathematical notations around them and on the back of the papers. His table lamp threw a bright pool of light in the corner of the dark, shadowy workshop. The night was completely silent, save for the distant sighing of the wind through the pines outside, the faint scratching sound of his pencil, and the clicking and whispering of the slide-rule in his hands when he paused to compute some factor in the diagrams.

Building and weight-testing the gravitors that went into the ship had led to speculation of other possible uses of the mechanisms. The possibilities were many, and Morrow spent his spare-time working them out. His ability, however, was limited.

First, there was the electronic efficiency of the gravitors, the increased power gained from battery storage-cells, the decreased loss of power within the circuits and mechanism. If electrons worked more efficiently in a gravitor's field, then mechanical and chemical power might work just as well. It appeared, on paper, that a small, one-horsepower gasoline engine might deliver the equivalent of a hundred horsepower or more in electrical energy, if it were incorporated into a gravitor field. Morrow worked this "gravitor engine" out the best he could, cursing his lack of knowledge in mechanical engineering. It might work, but he didn't have the knowledge to tell exactly how it could be made. It wasn't his line.

Then, there was the possibility of using the increased gravitor-field efficiency in radio communications. This was right up his alley, but the implications went so far and so deep that only a thoroughly experienced and trained scientist could trace all of them. He hadn't been an engineer long enough to have acquired that much training and experience; he wasn't a renowned scientist in the field. He couldn't always be sure where he was right or wrong in his computations. This was pure research; no book had ever been written for it. He couldn't look up all the answers.

But it appeared that a small radio set would have the power to reach anywhere in the Solar System, not to mention the extensive refinements of any television and/or radar set-up.

The possible refinements of chemical catalysts and electro-chemical processes were extensive, too. Staring at his diagrammatical results, Morrow wondered if mechanisms couldn't be perfected to measure the taste of foodstuffs as the taste-buds in the human mouth did, to measure the smell of odors as the human nose did, to convert carbon-dioxide into oxygen as plants did—even mechanisms which would react selectively to the electrical impulses generated by the cells of the human brain!

But he wasn't a chemist. He could only guess at the possibilities.

Finally, there was the possibility of applying the gravitors directly to the problem of transporting the human body by air. Part of this, he could answer: a gravitor strapped to a man's back would more than replace the conventional parachute for emergency bail-outs. The gravitor could be hooked into alternate power-circuits with alternate field-transmission coils, so if it failed to work on one setting the wearer could switch it to another, the equivalent of wearing a second parachute in case the first failed to open. And unlike a parachute, the wearer would have complete control over his rate of fall: he could descend gently to the ground or, if he wished, he could stop and hover in the air or even reverse his descent and rise upward.

That was part of it. Morrow had discussed it with Smitty and they'd decided to incorporate it into their project. In addition to having the ship look like something from outer space, there was also the problem of having to make a forced-landing somewhere. They might be seen on the ground, repairing their ship. The gravitors could be built into a tank carried on their backs, and fastened to a special harness costume complete with transparent helmet fitting over their heads. The helmets would protect their faces from the wind in a bail-out. Also, their appearance would be altered just enough to make them seem to be visitors from another planet, beings who did not breathe Earth's atmosphere.

But that still didn't give the human body a means of transportation by air. A small, portable propulsion unit was needed for that, and Morrow wasn't at all sure he could design such a unit. He was not a jet engineer.

He wasn't too sure about the large propulsion unit in the tail of the ship, either. Basically, it was a ram-jet unit. It ought to work, but it might not work too well....

Morrow tossed down his pencil and slide-rule, sighing, then pressed his hands over his aching eyes and rose from the table. *It's too much for one man!* he thought bitterly, and dropped his hands to his sides.

He stood gazing into the deep gloom of the workshop, at the huge, black hull gleaming softly in the darkness. Fifty-five feet long and fifteen feet high, the ship rested patiently on the narrow runners that supported its sleek belly. Twenty-five hundred dollars and six weeks of cautious, painstaking work rolled into one beautiful, fantastic-looking black monster with curved fins around the cluster of "rocket" tubes in its tail and streamlined, submarine-type diving vanes near its nose. Those vanes had been Smitty's contribution, operating on a cross-control system to bank the ship and lift it around a turn as the aileron-elevators did on flying wing aircraft. No other control surfaces were installed; the long, sleek rudder fin was immovable.

The night wind soughed through the forest on some nearby mountain slope. The ship stood black and silent, gleaming softly in the deep gloom of the workshop. It was a weirdly beautiful thing, like some creature of the Unknown.

*Straight out of the science-fiction magazines!* Morrow mused, grinning. *If Gwyn could only see it—*

A vision of her rose into his thoughts: Gwyn, lying on her stomach, the tight roll of her swimming trunks about her thighs, the smooth, tanned skin of her slender body, the firm swell of a breast beneath her armpit, the sunlight glints on her brown hair and the cool, calm wariness in her eyes....

Morrow grimaced wryly. Gwyn again! He'd been thinking entirely too often of her, and too much, since he'd left Westerton. He kept telling himself she was just another of the sacrifices he'd been forced to make, another part of his life he'd had to deny himself—

Still, when he slept he dreamed.

He was just too damned young, he told himself harshly. The demands of his body were strongest at his age; it wouldn't let him alone. His instinct to mate, to reproduce his kind, demanded satisfaction. There was danger in that. If he fought it, denied it, kept it bottled up inside him, it could spread and infest his whole being until it became a perverse fixation on sex. He had to have some outlet for it. Time off from his work, time to relax and enjoy female companionship, the nearness of a woman. An older man, in whom the mating lust had had time to diminish until it wasn't quite so strong and insistent—an older man could retire and live in an ivory tower of science. He couldn't. He must make allowance for it.

Find himself a girl in town. A date, a little moonlight and soft talk. Forget about a girl three thousand miles away. Forget Gwyn....

But he wished she were here. He wished she could see the ship.

Dawn was etching its rose-colored light in the East when Smitty drove in the yard.

They installed the batteries and climbed out through the simulated air-lock entrance to the ship, peeling off their gloves and shoving them into their hip pockets. Smitty turned, wiping his hands on his coveralls, and looked up at the ship.

"We can ground-test her without taking her outside," he said plaintively.

Morrow picked up his mackinaw and slung it over his shoulder, grinning. "Can't you wait 'til tonight?"

Smitty scowled at him. "Suppose she doesn't check out? Then we'll spend the rest of the night overhauling her! We oughta give her a ground-test right in here, Bill."

"Fair enough—if she doesn't go through the roof! But let's wait 'till after breakfast, anyway." He walked over to the stove, checked its fire, and shoved a couple more sticks onto it to keep it burning until they got back. "C'mon," he prompted, heading for the door. "I'm hungry if you aren't!"

They left the workshop and crunched through the brittle ground-frost to their shack. Morrow took his turn as cook, whipped up a batch of sausage, eggs, and pancakes, and boiled the coffee to the strength he preferred—which Smitty diluted liberally with canned milk. They gulped down their breakfast, cleaned the dishes, and strode deliberately back to the workshop. The chill November air bit into their clothes, but neither hastened his pace.

As they entered the warmth and shadow of the workshop and pulled off their coats, Morrow felt a fluttery sensation in his stomach which he carefully neglected to mention. It was probably indigestion, anyway.

Smitty, too, was silent. He tossed his coat on the workbench, strode straight to the open air-lock door, and clambored up into the ship. A tight grin creased Morrow's face as he followed with what casualness he could muster.

They moved through the luxurious forward lounge and climbed the metal steps into the control pit. Smitty slipped into the pilot's seat behind the controls and flight panel, up forward. Morrow took the flight engineer's seat behind the instrument console, on the left side of the transparent blister dome. The console sloped gently, like a desk-top, its surface glittering with a dozen instrument dials, twenty-four switches, forty-eight signal lights, two knobs and master switches, and a jet-blast temperature gauge.

"Flight station checks," Smitty reported quietly.

"Roger." Morrow swept his hands across the console, flipping on the twenty-four switches. "Stand by for gravitor check," he added, then clicked on the two knobs.

The ship shifted slightly beneath them. The faint, sighing sound of wind came from the tail.

On the console, twenty-one signal lights flashed blue. Three flashed red. Morrow scowled at them.

"Report gravitor check!" Smitty prompted impatiently.

"Three gravitors out," Morrow growled. "One auxiliary lifter, one auxiliary and one main drive gravitor. Must be a short in 'em somewhere."

"We don't need the drives for a ground-test," Smith reminded him. "Cut to main lift units and let's try her out!"

"Wilco." He switched off the drive knob and the twelve auxiliary lifter switches. "Stand by to rise!"

The sighing wind was gone from the tail. He gripped the lifter knob in his fingertips and, turning his head, stared out at the dark floor of the workshop below.

He turned the knob, cautiously.

The ship rocked gently, then lifted. The floor dropped away beneath them.

"Watch it!" Smitty warned tersely.

The ship paused, then seemed to settle.

They floated serenely, twenty feet above the workshop floor. The heavy rafters of the roof loomed close over the transparent blister.

Smitty cleared his throat, nervously. "I think that's high enough!" he exclaimed.

Morrow permitted himself a fleeting grin, then began to inch the knob back toward its stop. "Stand by for descent!" he warned.

The ship settled slowly. The floor rose up with majestic deliberation—then paused again.

"How high are we?" Morrow asked.

"A little over four feet on the altimeter," Smitty replied. "Want to hold her here a while?"

"I want you to climb out and see how much it alters her lift," Morrow explained. "One less passenger shouldn't affect it at all, but let's make sure."

"Wilco." Smitty rose from his seat and came back toward the steps.

"Jump around a little," Morrow said. "See if it rocks her any."

Grinning, Smitty banged noisily down the steps and clattered back through the ship. She rode perfectly still, unmoving. Smiling his satisfaction, Morrow waited.

Then Smitty walked around the bulge of the nose, on the floor below, and waved to him. Morrow waved back and, rising, moved up to the front seat. The altimeter still registered slightly over four feet. He returned to the console, sat down—and snapped off the lift knob.

The ship settled immediately to the floor, struck lightly, and rocked to a standstill. Morrow clambored down the steps and felt his way back through the dark interior to the air-lock.

Smitty was waiting for him as he dropped to the floor. "She checks, doesn't she?"

"She checks," Morrow affirmed. "Now let's get to work on those shorted gravitors!"

The first night's tests were preliminary. They lifted the ship a few feet off the ground and flew it across the sawmill yard and back. They switched the gravitors from main to auxiliary systems. They loaded the cargo deck amidships with sandbags and flew a weight-test. They took the ship up to fifty feet and held it there until the wind, blowing them toward the trees, forced them to come down.

The ship checked out in every test. They decided to make the first trial flight the next night.

Morrow sat up in the co-pilot's seat beside Smitty as they drove steadily through the darkness. Above, the stars twinkled coldly in the black heavens and the white sickle of a quarter-moon threw its milky glow into the control pit. Below, rolling gray stretches of meadow spread out between dark, timber-clad shoulders and humps of the Sierras. To the East, the timber gave way to rocky, cloud-wreathed peaks. They were headed toward them, and climbing.

"Five thousand on the altimeter," Smitty remarked flatly. "That's ten thousand, five hundred above sea-level. She isn't levelling out yet." His face was grim in the green glow of the instrument dials.

Behind them, the black, glinting hull was crammed with sandbags. They were lifting a full load.

Morrow kept his gaze fixed on the air-speed indicator. A deep, whooming sound came from the tail-jets. The needle on the indicator dial flickered restlessly, back and forth, over a single point.

They were doing forty miles an hour, indicated air-speed. They hadn't been able to increase that speed. A brisk twenty-five-mile-an-hour wind was blowing them steadily southward off their course.

Smitty shook his head. "Those jets don't even *sound* right, Bill—"

"I know," Morrow said. He sighed wearily. "We've got to do better than this. Take her higher—ram-jets are supposed to work better at high altitude."

"I don't want to go over twelve thousand without oxygen," Smitty replied. "Can't let this wind blow us down over Yosemite National Park, either—if we can help it."

"Take her up," Morrow said.

The ship continued to rise, steadily.

"Eleven thousand," Smitty chanted. "Eleven thousand five hundred, twelve thousand, twelve thousand five—she's flattening out!"

Their ascent slowed, gradually. The ship steadied at thirteen thousand feet above sea-level—7500 feet on the altimeter, which had been zeroed to the altitude of their sawmill workshop.

"Down!" Morrow barked. "She's losing speed!"

The indicator needle was creeping back past thirty-five, then thirty— their sideward shift to the south could be felt. Smitty shoved forward his control wheel. The ship dived.

They glided easily back across the mountain slope toward their sawmill. Judging their wind-drift accurately, Smitty set the ship down in the yard before the black, yawning doors of the building. As the runners scraped the ground, he switched off the gravitors and slumped back in his seat, dejectedly.

"We've got to rebuild that jet chamber," he muttered. "There's something wrong with it, Bill. All we've got is a big wind-blower, in spite of her weightlessness—the drag of the hull wouldn't slow us down *that* much!"

Morrow unbuckled his seat-belt, rose, and strode back to the steps without a word.

It took them a week to pull out the rear bulk-heads and completely redesign and reconstruct the tail-jet assembly. When they finished, they tried it again. They got an air-speed of seventy m.p.h. at low level, but it dropped to twenty m.p.h. as they gained altitude. The tail-jets didn't just make a whooming sound, this time—they made a rumbling, burbling sound.

They landed and pulled the ship into the workshop, closing the big doors after it. Morrow walked over to the workbench, pulled off his gloves, and threw them down.

"It's no good!" he said harshly. "That jet chamber just isn't shaped right—there's too much turbulence in it, breaks up the jet-blast."

"We'll rebuild it again," Smitty said, with a shrug in his voice.

Morrow wheeled and glared at him, red-eyed. "We aren't jet engineers, Smitty. We're building by guesswork! We can redesign that jet chamber a thousand times and never get the right shape!"

Smitty moved on to the stove and began stoking up the red coals, stacking wood on them. "She does seventy per hour up to seven thousand feet," he said dully. "If that's the best we can do, we'll just have to be satisfied with it."

"It's not *good* enough!" Morrow protested. "She *has* to have more speed, Smitty. She'll be at the mercy of every wind that comes along if she hasn't, weightless as she is!" He smacked his fist into his palm, decisively. "We've got to get help, chum."

"Help?" Smitty turned and looked at him, querulously. "Where can we get help?"

"A jet engineer," Morrow snapped irritably. "That's the only one who *can* help us. We've got to find one—" He broke off, suddenly thoughtful.

Smitty grinned without mirth, mistaking his silence for hopelessness. "That's the trouble, Bill," he said. "There's no one who *would* help us!"

"I'm not so sure about that!" Morrow replied softly. "*I'm not so sure at all*—"

It was late Friday afternoon when Morrow parked the battered, mud-splattered truck on a side-street and climbed out to go for a quiet stroll in suburban Sacramento.

The street address he was looking for turned up in the next block, near the edge of town. It was an inconspicuous one of the long street-row of small houses with a green lawn stretching down to the curb and dotted with a few evergreen shrubs, a broad livingroom picture window in front, a white front door with a small ornamental iron night-lamp mounted above it, and a one-story, red-tiled roof in the flat, gently sloping California style.

Morrow walked past the house and around the block to the alley. He walked up the alley behind the house. Its rear was as inconspicuous as its front: a wide yard, partly in lawn, partly in flower garden and part gravelled with clothesline, enclosed by a low, whitewashed wood fence. The only noticeable difference was a small sand-box in which a small brother and sister were playing with toy cars. The little boy and girl wore matching rompers and had straw-colored hair which, Morrow concluded, they must have inherited from their mother. He'd never met Mrs. Foster, but he remembered Bob Foster's dark, heavy hair.

He walked on down the alley, studying the back yards behind the other houses. He noted how wide the alleyway was, how high surrounding fences, garages, and other obstructions were, and the lack of telephone poles or wires overhead. He nodded his satisfaction.

When he got back to the truck, he took a street-map from the glove compartment and carefully marked the exact location of Foster's house.

Then he drove out of Sacramento, had dinner at a roadside restaurant, and proceeded to Stockton. Smitty met him downtown and they went into a lunchroom for coffee.

"Groceries and laundry's taken care of," Smitty reported wryly. "How was Sacramento?"

"Fine," Morrow said. "If the weather forecasts for tomorrow night pan out, we'll get in and out without any trouble."

Smitty frowned worriedly. "It's still a big risk to take, Bill. We'll be flying into the Coastal Radar-Defense Zone, you know, and we can't just file a flight-plan at an airport for an unauthorized, illegal ship. I'd hate to look up and see an F-140 night-fighter with its nose-cannon blazing at me!"

"That ground radar isn't effective below three thousand feet," Morrow reminded him. "I think we can sneak in at treetop-level without being detected."

"That's all right, unless we fly into a power-line in the dark," Smitty grumbled. "It's still risky as hell—"

"We've got to have Foster," Morrow said firmly. "I can't say for sure whether he'll join us or not, but we've got to try!"

"Okay!" Smitty signed resignedly. "We'll try."

The following night, Morrow left Smitty checking over their ship and flight equipment and drove the truck down to a gas station on the highway, thirty miles west of their sawmill-workshop. He parked beside the gas pumps, told the attendant to fill the tank and check the oil, and went inside to the pay-phone booth.

He called Sacramento Long Distance and gave them Foster's home videophone number.

There was some fault in calling from a pay-phone, of course—and a Long Distance call on a rural pay-phone at that. Neither Long Distance calls nor pay-phones nor rural phones had the new videophone accessories. Videophones, involving two-way television transmission via a camera-screen installation, were still in the development stage. Metropolitan and suburban phones had the video screens. Long Distance coaxial transmission was still too costly to merit the installation of the screens on rural phones—which also ruled out Long Distance video calls. To install the screens in pay-phones would, as yet, triple the cost of the calls.

Naturally, the Sacramento operator would inform Foster this was a Long Distance call; Foster's screen would remain blank. The gas station's pay-phone had no screen. This was a disadvantage to Morrow: not seeing

Foster's face, he wouldn't be absolutely certain he was speaking to Foster. He'd have to rely on his memory of Foster's voice, and it had been more than two years since he'd met Foster.

Positive identification could be important. Morrow kicked himself mentally for not making a local call to Foster's home while he was in Sacramento. Suppose Foster had moved in the past two years? Suppose there was some sort of slip-up that aroused someone's suspicions just enough to start the authorities on an investigation—

Even the *slightest* mistake might finish them!

And the call had to be made. Their plan was set for tonight, Saturday night, because Foster was most likely to be home from work—research engineers often worked late hours on weekdays—and because he'd probably have the next day off. They had to get Foster out for that one day, and it had to be done right. But they had to be certain that Foster was home when they went after him.

The receiver continued its rattling noise in his ear as Morrow waited, fidgeting impatiently, and the seconds crawled past.

The rattling ended with a faint click.

*"Hello?"*

Morrow exhaled a shuddery sigh of relief. He recognized Foster's characteristic deep, muffled tones almost at once. "Hi, Bob. This is Bill Morrow—"

*"Morrow? Well, hi yourself! Where you calling from?"*

"I'm on the highway," Morrow said. "I'm on my way north and wondered if I might drop in as I pass through Sacramento—I ought to be there in a few hours. You going to be home?"

*"Ye-e-es. C'mon around, by all means! You still have my home address, haven't you?"*

"Sure thing. How've you been?"

*"So-so, between drawing curves on flight-test characteristics and pounding out stories. You written anything lately?"*

"I've been a little too busy to give it much thought," Morrow answered truthfully.

*"Uh huh! Well—say, you going to be in 'Frisco for next year's science-fiction convention?"*

Morrow grinned. "'Sa little too early to say, yet. I'll see you in a few hours, then, huh?"

*"Right-o! We'll have the beer on ice!"*

Morrow drove back to the sawmill workshop and helped Smitty perform a final inspection of the ship and equipment. Their plan was worked out thoroughly. The ship would fly to and from their target at low altitude, and at its maximum speed. The forecast weather conditions would aid in hiding them, but it would also hinder their flight—much of it would have to be done on instruments, and Smitty spent considerable time studying topographical sector-maps and radio omni-range vectors.

Their personal gear consisted of two special suits which would serve to conceal their identity as well as aid them in an emergency. The suits, patterned out of shimmering fabrilastex material, fit with skin-tight snugness over their long winter underwear and socks. The foot-soles of the suits were of springy foam-rubber, heat-welded to the fabrilastex just as the seams in the material were heat-welded to a perfect fit. A sturdy harness fitted into the inside of the suits to grip their legs, thighs, and chests, suspending them in bail-outs from the sturdy plastic tanks on the back of their suits. Each tank enclosed a gravitor unit. A lightweight, transparent blue dome helmet fitted over their heads and clamped onto fasteners on their shoulders. There were small air-vents around the bottom of the helmets and in the fantastic-looking knob attachments in their tops.

They pulled on their suits in the workshop and stared at each other, grinning. "All you need," Smitty taunted, "is a flashlight ray-gun in each hand!"

"You look pretty monstrous yourself, blue-face!" Morrow retorted.

"*You* look *sexy*, old boy!"

"*Down*, Rover! Better climb on the ship's radio and check the weather reports again—"

"Wilco!"

Morrow walked to the end of the workshop and swung open the big doors. Then he went back and crawled into the ship, swinging the thick "air-lock" door into its grooves behind him. As he climbed into the control pit, Smitty reported that the weather was just as lousy as they wanted it to be: clear, cold, and windy at high altitudes, with some low cumulus and a five-hundred-foot thick blanket of fog hugging the ground and creeping in and out of the valleys. There were several scattered thunder-showers and by morning there would be solid rain in the mountains.

Morrow switched on the gravitor units at the flight engineer's panel, then moved up and strapped himself into the co-pilot's seat. "It's your bus, Junior," he said. "Let me know when we reach my stop."

"Passengers move to the rear, please," Smitty retorted, and eased the ship cautiously out of the workshop. They swung northward and set off, flying just a few hundred feet above the mountain slopes. The moon was a cold, white gash in the black heavens, and the dark mantle of the treetops swept past below.

Unfastening his helmet, Morrow swung it back and relaxed, lighting a cigarette....

They had to use every precaution in going after Foster. In the first place, they had to consider that he might be violently opposed to their project—that, in fact, he might go straight to the authorities with it. The only safeguard against that was simply to prevent Foster from knowing where their project was located. Without that information, he would probably find it difficult to make the authorities believe him. A mere story about mechanisms that control gravity, without any basis of fact to support it, would sound rather far-fetched.

For that matter, it would have been difficult merely to visit Foster and convince him they did have such mechanisms! The only quick answer was to show him, to prove it to him. Then he would listen to them.

There was a good chance that he'd approve of their project and help them with it—otherwise, Morrow wouldn't have thought of him. And he was a man who could help them. Robert Foster was a jet engineer, employed as a flight-test analyst at an aircraft corporation's experimental plant near Sacramento. Morrow had met him, however, because Foster had written many stories for the science-fiction magazines, mostly on the galactic empire theme. They had met at a private science-fiction club in New York and spent most of a long night in a bar, along with several other writers and magazine editors, discussing subjects of vast scope and consuming beverages in vast quantity. Foster had proved himself a kindred soul of fertile imagination, if not of superior intellect, and so into the wee, small hours.

In short, Foster had impressed him as a man to be trusted when the going got rough.

Whether or not that impression had been correct, Morrow didn't know. Tonight would certainly put it to the test. They could only ask for his help, and that was all. If he refused, he refused. They couldn't use threats or coercion or any suggestion of violence—that would gain them nothing.

Foster *had* to agree! There was no one else! Without his help, they were stymied....

The weather thickened as they turned west, coming down off the slopes of the Sierras. Silvery masses of cloud drifted by in the moonlight and a thin, gray haze obscured the ground. They cruised along, their tail-jets rumbling, descending slowly to pass beneath a long row of clouds ahead. Raindrops began streaking the transparent blister which pinged at their impact; then it began a steady, ringing sound as the downpour increased. The world was turned into a gray, trickling wetness, faintly reflecting the green glow of the luminous instrument dials. The lights of a town appeared off to the left, wavering sparks in the wet gloom. Smitty swore under his breath.

They emerged from the shower to find themselves over an endless mass of cottony white, completely hiding the ground. "Now we gotta go down through that stuff!" Smitty muttered, and pushed the nose down.

The ground became dimly visible through the mist at a height of seventy feet. "Airspeed's a hundred and ten; headwind was reported at twenty miles." Smitty chanted glumly.

Morrow said nothing for a moment, knowing Smitty meant that if they were flying any faster their dim, wavering view of the ground would mean nothing. Then he started and looked up. "A hundred and ten? In a twenty-mile wind? That's ninety miles an hour!"

Smitty stared at his instruments and nodded slowly. "We're doing better than we did," he agreed. "Either that, or this wind has twisted its tail. We'll check it again."

They flew onward through the swirling, dark mist. The dark blurs of trees flashed past below, and houses, roads, and telephone lines. Dim, shadowy objects, hardly recognizable. And there were moments when the mist closed in completely, hiding everything. Morrow felt a cold sweat forming on his face. The jets made a deep, mournful rumbling sound in the ship's tail. A highway swept past below, with car headlights revealed as moving blobs of yellow in the darkness.

"This is the block," Morrow said, finally. "Swing across it and come down in that alleyway in its center. I'll tell you where to land then."

Below them were the familiar rooftops of the houses, rising darkly out of a thin ground mist. Smitty brought the ship over them, cutting the jets, and let it coast to a stop over the narrow, vague band of the alleyway. Slowly, they drifted downward.

Morrow consulted the street-map on his lap again. "Up a little further," he directed.

The jets gave a brief, rumbling sigh and they glided forward.

"Here—ground her!"

Gravel rasped against the ship's belly. They unfastened their belts and scrambled down into the ship.

"What time is it?" Smitty whispered, as Morrow swung open the door.

Morrow glanced at his wrist-watch. "Three-ten a.m.," he said half-humorously. He wondered if Foster was still waiting up for him. "Fasten your helmet down, and let's go!"

They dropped down from the ship and went over to the low, white fence behind Foster's house. Passing through the gate, they strode across the yard. The mist-shine glimmered faintly off their bodies. Their blue-tinted helmets were grotesque globes of darkness, like the heads of nightmare creatures.

Light glowed from a window in the side of the house. "Somebody's up!" Morrow observed softly.

"Do we go 'round and ring the front doorbell?" Smitty wondered. "Or do we just walk in?"

Morrow shrugged. "It won't make much difference. Let's try the back door—if it's locked, well go around."

They reached the door and he tested its knob, careful not to make any noise. It yielded readily.

They entered.

The faint light filtering down the short hallway was enough to guide them across the dark kitchen. Then they had to pass the dark doorways of what were probably two bedrooms, on either side of the hall. They reached the lighted doorway near the front, and stood looking into the living room.

Robert Foster was seated in a comfortable chair next to the television set. A single reading lamp was burning—the pipe clutched in Foster's teeth was out—and he seemed deeply engrossed in a good book.

Morrow reached up and snapped the fasteners on his helmet.

Foster lifted his gaze with the utmost casualness and studied the two figures in the doorway. He looked quite happy and contented, dressed in an old pair of slacks and loafers and a turtle-neck sweater. His dark, touselled hair showed evidence of his hand running through it—a habitual gesture of his, Morrow remembered.

Slowly, a stunned expression crept across his face.

Morrow swung his helmet back onto his gravitor tank. "Hello, Bob," he said.

Foster slipped a marker into his book, closed it, and laid it carefully aside. "Morrow?" he said. "So you finally made it! I might've known you'd be coming by way of Jupiter—but why the get-up, friend? And who's your partner?" There was just the slightest quaver in his voice.

It was almost more than Morrow had hoped for. He could play it through, now. "This is a Martian friend of mine," he said, hooking his thumb toward Smitty. "I can't stay long. Somebody might see our spaceship and get curious."

"Your—spaceship?" Foster queried falteringly.

"We landed it out in back."

The room was silent for a moment. Foster sat dumbfounded, staring at them. A flicker of a gleam began to show itself in his eyes. "Am I to understand," he said gently, "that you have landed a spaceship in my back yard?"

"No," Morrow corrected. "In the alley."

"Hmmm—it'd better be in the alley. My wife would slaughter us both if you'd trampled her gardenias." Foster fell back in his chair. He tried to relax; he even grinned, somewhat shakily. "Now what's the idea, Bill? Why'd you come tippy-toe in here like this? Out with it!"

"Take too long to explain," Morrow replied, shaking his head. "Somebody's liable to see that spaceship any minute, now." He forced a broad, innocent grin across his face. "You want to come have a look at it?"

"Ye-e-es!" Foster agreed sarcastically, rising from his chair. "I suppose I *should* take a look at it—"

Morrow led him out the front door and around the house. "Don't want to awaken your wife," he explained, clamping down his helmet.

"No-o-o-o!" Foster conceded. "I wouldn't advise that!" They proceeded on across the back yard, through the clinging, wet fingers of the mist.

Then Foster saw the ship.

After that, it wasn't too hard to persuade him to enter it. Then it was simple to switch on the gravitors and rise into the dark sky. Morrow had him planted in the flight engineer's seat, enthusiastically demanding explanations in full, as Smitty piloted them swiftly homeward.

Foster was sold!

They held a conference in the sawmill-workshop that lasted all the next day and well into the next night. Then Foster went home to tell his wife he'd had a hurry-up call from the aircraft plant and gone there to work on some secret research; they drove him back to Sacramento in the truck, and let him off near his house.

Then they returned to the workshop and went to work.

The following weekend, Foster drove up in his own car to see them. He climbed out of his car wearing lace-boots and hunting clothes. Reaching into the back seat, he brought out a shotgun and a stack of newspapers, then Morrow came up to greet him and they strode into the workshop.

"You fellows have really been hitting the ball!" Foster exclaimed, as he stopped and gazed at the small, needle-nosed ship sitting beside the larger ship.

Morrow nodded. They had worked night and day to construct the second, smaller ship—a little two-passenger job with sweptback fins and a canopy-covered cockpit in its sharp nose. It rested neatly on its long A-fins, poised to hurtle into the sky. Its color scheme—dark blue-black on top, light gray on its belly—stood out in sharp contrast to the solid, shimmering black of the giant ship behind it.

It had been Foster's idea. He'd pointed out to them that they needed a smaller experimental model, easier to dismantle and rebuild, for the development of their air-jet chamber.

"Have you given it a test-flight yet?" Foster asked.

"Ran it out last night," Smitty replied, coming around the two ships to meet them. He set a plumber's blowtorch on the workbench and wiped his hands on a rag. "It hit seventy miles an hour, then worked up to seventy-four after a five-hour run."

Foster shook his head in puzzlement. "That's something I just can't account for. A jet-pod ought to be just as efficient as its design, and nothing should alter its basic performance other than a change in atmospheric conditions."

"There was no atmospheric change," Morrow said. "Same altitude, same barometric pressure, same thermal conditions. I'm beginning to think the problem isn't only in the jet-pod design."

"That makes two of us!" Foster agreed. "The design I gave you should've worked better than any seventy miles an hour, if your propulsion unit develops that focus of 'false gravity' and squeezes the air out, forming a low-pressure center, as you said it did."

"We've checked that, too," Morrow said, frowning thoughtfully. "I'm beginning to think it's something to do with the gravitors' field of influence. Come over here—I want to show you something!"

He led the jet engineer over to where he and Smitty had rigged a gravitor mechanism and a sling-load of sandbags with rope attached, just as they'd used in weight-testing the gravitors. He switched on the gravitor, adjusted its setting, and let it lift the load of sandbags into the air. Then he pointed to the rope dangling down beneath it.

"See that twist in the rope, just under the sandbags?" he said. "That much of the rope is in the influence of the gravitor's field, which is cancelling out the pull of the Earth's gravity. Now then, if it can influence that three-foot length of rope, what influence might it have on the air around it—and on the slipstream of air flowing over our ships, which is supposed to enter the air-vents and be blasted out the jets for propulsion of the ships?"

"It could be scrambling our intake flow," Foster acknowledged pensively. "But would that condition alter in time?"

Morrow shook his head. "I don't think it does—or that it would unless the gravitor's batteries were almost burned out. Then the field's influence might lessen a bit. Otherwise, no."

"Then why is it that the jets' efficiency increases with time?" Foster asked. "How'd you get seventy miles an hour on the big ship, then ninety? And five hours' running built up the little ship's speed an additional four miles per hour, didn't it?"

Smitty nodded. "It gets gradually better—but not much. If we knew how it happened and what it was doing to the air-flow, maybe we could design jet-pods with the right shape to use that air-flow and get good performance."

Foster turned and peered sharply at Morrow. "Bill, doesn't that gravitor's field work by conductivity of some sort through the surrounding material?"

"Uh?" Morrow started. "Yes, it—wait! You mean the ship's plastic hull?"

"Right. And what about the polarization of that plastic?"

Morrow pursed his lips, contemplatively. "Like all materials on Earth, it's polarized—if you want to use that word—to the gravitational and magnetic fields of Earth. I see what you're driving at, though—the gravitors establish a field in which the Earth's gravity and magnetism are cancelled out, or bent back upon themselves. The mechanism of the gravitors, the hull they support, everything within their field of influence is placed on a basis of its own gravity, mass-attraction, magnetism, what-have-you."

"And that's gradually changing the polarization of those materials," Foster concluded. "And the gravitors' field, working through the material, is also affected. There's a gradual change in its influence on other surrounding matter—and on the slipstream flowing over the ship!"

"We'd need a wind-tunnel to test that, wouldn't we?" Smitty asked dejectedly.

"Yep," Foster agreed. "And wind-tunnels cost money. The only other way to test it would be to make a cross-country flight, and I wouldn't advise that."

"What about a cross-country night flight?" Morrow wondered.

Foster gave him a strange look. "You two haven't been reading the newspapers lately, have you?"

Morrow and Smitty exchanged glances of mingled surprise and guilt. "We've been rather busy out here," Morrow protested lamely.

"I suspected you were," Foster said, a trace of grim humor in his voice. He walked over to the drafting table in the corner, where he'd left his shotgun and bundle of newspapers. "Pull that thing down and come over here," he told them. "I've something to show *you*, now!"

Morrow cranked the gravitor-sling down on the hand winch and Smitty shut it off; then they went over to where Foster was spreading newspapers on the drafting table, checking and circling columns of newsprint with a blue crayon pencil. Morrow stepped to his side and stared down at the papers. The words fairly leaped up to strike him in the eye.

**MYSTERY SHIP NEAR SACRAMENTO**

**BLACK SPACESHIP SEEN**

**MARTIANS PREFER CALIFORNIA!**

**TWO CARS LEAVE H'WAY AS ROCKET SWOOPS**

**BLACK ROCKET SHIP; 'NOT OURS,' SAY AIR FORCE**

There were more than a dozen news stories about it—not front-page, black-headlines stories, but two-column stories beginning on page two or three and continued in the newspaper center-section. None of it was spectacular enough to merit big headlines.

However, it had obviously been given a thorough coverage by the press. A railroad worker walking to work the Saturday morning of their trip to Sacramento had seen "a black, torpedo-shaped ship flying through the mist at low altitude, making a deep, rumbling noise." A police patrol car on the highway had seen it "flying low through the clouds, as if it were having mechanical difficulty of some sort." Two cars had left the highway and skidded into a ditch as both drivers saw "a black ship without wings swoop directly over" with a sound "like one long, continuous A-bomb explosion!"

Some said the ship was just a solid black shape, without lights or any noticeable features except the absence of any wings; some said "a long, blue flame" came from the tail of the ship. Some said "bright red, green, and blue lights were swarming around it" and some claimed there were "big windows in the sides, with something moving around inside."

Officials of the Air Force, both in California and in Washington, professed to have no knowledge about the ship. But one fact was added: both official groups said they were deeply interested in the reports for "reasons of security," that a thorough investigation would be made, and that radar surveillance along the West Coast would be intensified.

And one, final news story was headed: SEARCH FOR DOWNED 'SPACESHIP' FAILS. There had been strong belief, it said, that the mysterious black ship had been in trouble and was making a forced landing when it was sighted.

"There it is," Foster said with a tone of finality. "These are all the stories in the local papers. It's been played up from coast-to-coast, however—both in the newspapers and news telecasts. And the defense forces along the Coast are just waiting for you to pop out again so they can pounce on you."

"Along the Coast," Smitty echoed pensively. "It's significant that they haven't turned their attention to the interior—back as far as the Sierras, here—"

"Probably think it's some sort of new Russian reconnaissance aircraft," Morrow interjected. "They undoubtedly have a nice, little reception committee waiting out over the ocean."

Smitty nodded. "Any cross-country we plan to do had best be plotted due east, across the desert."

"There's the atomic project area, that way," Foster protested. "They certainly must have increased their air defenses around that."

"At low altitude, we can get around it," Smitty said.

Foster's features went slack. "Look here! You're not seriously thinking of—"

"If we had a wind-tunnel, no!" Smitty retorted wryly. "We could stick the little ship in it, let it run for a few days, watch the hull polarize itself to the gravitors' field, and note how the air-flow around the ship was affected. Then we could rip out the jet chamber and design a new one that'd work in the affected air-flow."

"*If* we had a wind-tunnel," Morrow emphasized.

"Right!" Smitty turned back toward the ships. "So," he concluded, "we take the big ship! We head out over the desert and keep going, watching how the ship performs and what the air-flow does to her. We'll have to install a few barometric pressure-point indicators around her hull—"

"But we'd have to fly several days steady to get that hull completely polarized," Morrow said. "We can't just restrict ourselves to night flying."

Smitty winced. Then he rubbed his chin, scowling. "If we have to, Bill, we can go east to Utah, then south through Arizona to Mexico, then east again—flying across the Border at night, without lights, won't be too much trouble; and once in Mexico we won't have to worry about radar. We can go out over the Gulf of Mexico, if we want to, and then turn north and fly up the Mississippi and Ohio valleys as far as Pennsylvania. There's a lot of brush country in the neighboring mountain areas—there'd be little danger of getting seen through there. So long as we don't have to land anywhere, we're safe!"

"In other words, it'd be a cross-country endurance flight," Morrow surmised.

"But suppose the ship fails on you?" Foster demanded tersely. "Suppose you're forced down?"

"We're visitors from outer space!" Smitty replied, grinning.

Foster wasn't amused. "Let's not be foolish about this," he argued. "We've got something here that we can't let loose! The world isn't ready for it—"

"But we've got to have it perfected when the world *is* ready," Morrow said firmly. "Once the tension wears out and the world situation changes, we've got to act! If we aren't ready, the world will go right ahead and get mixed up in some other squabble. Then we'd have to wait again."

Smitty laid a hand on Foster's shoulder. "You can get a few days off from the plant, can't you?"

"What? Well, yes," Foster stammered. "Of course! But—"

They took off at noon on a cloudy winter day.

They spent the afternoon dividing their attention between the test-flight instruments and the surrounding sky. They hadn't the money to afford elaborate recording mechanisms to graph every moment of the flight onto neat tape-spools; they had to rely on the human eye, the questionably analytical human mind, and the servo-mechanism of a human hand wielding a pencil on a loose-leaf notebook. And they constantly expected to see a razor-winged jet fighter hurtling down from the stratosphere above them, its cannon sparkling the bright flame-color of death.

They didn't talk much that afternoon.

They took turns at the controls and eating until each had consumed his dinner, then gathered tensely in the control pit as the ship bored rumblingly into the black night. Ahead of them was the Mexican Border. Below them and around them, almost scraping the ship's belly, as low as they were, was the jumbled, boulder-strewn Arizona desert bathed in frosty white moonlight. Above were the cold, twinkling stars, the black heavens—and who could tell what radar-equipped night fighter poised above them, ready to peel off and plummet downward, guns blazing—

Then the Border was behind them. They took turns at the controls and instruments again, catching a few winks of sleep between turns. Morning dawned, and they approached the Gulf of Mexico.

Morrow checked their supplies—food and water for the trip, parts and materials stowed in the spacious cargo deck for repairs on the ship if necessary—and they took turns at breakfast. Then he and Foster sat down to an argument about the scientific implications of the gravitors. Foster was of the opinion that Einstein's theory no longer was valid, that Milne's work came closer to the truth but was still vague. Morrow thought differently, and they argued together amicably.

Noon passed, and they were over the green expanse of the Gulf. Smitty called their attention to the short-wave radio. The newscasts were quite interesting.

A professional hunter in Nevada, hired to exterminate a mountain lion which had been slaughtering a rancher's cattle, was surprised when a ship that looked "like a big, black whale" thundered over his head and plunged down behind a nearby ridge. The hunter rode hastily around the ridge, expecting to find the wreck, but the ship had vanished completely "as if the ground up and swallered it!"

A Greyhound bus proceeding across Arizona nearly swerved off the road when "a long, black torpedo at least a hundred feet long" came across the sky "so fast the air thunder-clapped behind it" and left "a trail of blue fire" behind it. Passengers on the bus verified the driver's story, with some minor variations.

Two farmers standing in a field in northern Nebraska saw a flight of six "fish-shaped" objects go over, each having a shadow "big as a barn" on the snow.

A noted banker in Chicago created an uproar when he reported seeing "a giant, black shape" rise from the waters of Lake Michigan as he was driving home in the afternoon.

An amateur astronomer in Alabama reported sighting a "strange ship" rising upward from the Earth's atmosphere "on a pillar of rocket fire." The ship had mysteriously disappeared "as soon as it left the atmosphere," the middle-aged hobbyist stated.

A Swedish Air Force jet-pilot claimed he had sighted, given chase, fired at, and seen his tracers bounce harmlessly off a "black, fish-like craft" flying at 40,000 feet above the Baltic Sea.

The news commentators added, in significant tones, that no airline pilots had yet reported seeing such craft. One added somewhat caustically that due to previous experiences the pilots probably wouldn't report anything to the authorities even if they did see anything, since the authorities persisted in treating such reports and the pilots who made them with painful ridicule; the commentator then launched into a condemnation of the current Administration.

"It would seem," Smitty observed from all this, "that we are quite famous!"

"'Notorious' is the word, I believe," Foster countered drily. "If this keeps up, some congressman is likely to introduce a bill providing that the government produce some Martians with black spaceships. The voters will demand it."

"It's good disguise for us, anyway," Morrow mused.

"Uh huh!" Foster grunted in reproof. "Unless we're found out, that is. If the public discovers that we've hoodwinked 'em and there aren't any Martian immigrants at all, they'll probably howl for our blood! I think this is going to develop into a scare-issue, Bill. I'm afraid people will want it, as an excuse to work off some of their nervous tension."

"Fine!" Smitty said grimly. "If anybody's trying to catch us, a general scare-issue will have 'em looking all over the place. We're already supposed to be in Nebraska, in Lake Michigan, in the Baltic Sea, and somewhere out in space!"

"Invisible, too!" Morrow laughed.

They passed over Louisiana in the early morning and proceeded northward up the Mississippi valley. Indicated air-speed was two hundred and thirty-eight miles per hour. Dawn was blanketed in a pouring rain. They turned off up the Ohio valley and reached the Allegheny Plateau in

West Virginia, flying by instruments, topographical maps, and radio omni-range navigation.

And once they almost blundered straight into a big, six-engined commercial stratoliner. The stratoliner pulled up almost at the last minute.

By mid-afternoon, they were approaching Pennsylvania. The drizzling rain had changed to snow and sleet. Then they were forced down. The ship's air-speed fell off with an alarming suddenness. Then the entire tail structure took on a heavy load of ice.

They settled tail-down into a clearing on a densely wooded slope. The ship wallowed deep into the soft, slushy snow.

The three men got together over the table in the forward lounge. Foster kept running his hands through his hair, nervously. "We're stuck," he said. "We're stuck here for the winter unless we can rebuild the tail assembly. That jet chamber has to be changed."

It was obvious, after they had diagrammed the readings from their various flight-test instruments. The ship's hull had become completely polarized to the gravitors' field; the field influenced the air flowing over the hull, so much so that a simple air-scoop couldn't pick up air to blow through the propulsion unit and out the tail-jets. The air intake had to be designed to work on the disturbed air-flow.

"It's a little like those 'space-warps' in science-fiction yarns," Foster explained. "There's a warp of the gravitational and magnetic fields around the ship. The air-flow entering that warp bends and twists to follow it."

"We ought to redesign the entire hull to comply with that warped air-flow," Smitty suggested absently.

"The hull doesn't matter so much," Foster contradicted. "We could design it in any shape, though a sharp nose and thin guide-fins are still effective. You just happened to hit the right answer when you placed the control-surfaces forward on the nose of the ship."

"Talking isn't going to get us out of here," Morrow remarked grimly. "Let's get to work on that tail assembly."

"I got news for you!" Smitty muttered. "If we rebuild the tail with our power-tools, it'll use up the juice in our batteries. We won't have enough to get home."

"We must get our batteries recharged, then," Morrow said. "Will we have enough juice left to get out of here when we're finished?"

Smitty nodded. "And then we'll be up a creek. Where do we get our batteries recharged?"

"Couldn't one of us venture into a town around here and buy a few batteries?" Foster suggested. "Without wearing our Martian costumes, of course."

"Our Martian costumes as you call 'em are at least warm!" Smitty retorted. "It's a little cold to go wandering around out there in our coveralls."

"Wouldn't pay to risk it, anyway," Morrow said. "Suppose someone has seen our ship flying around here? Suppose they make a report that brings in the authorities and—"

"But who'd think a man in coveralls just stepped off a spaceship?" Foster persisted.

"Uh huh. You have a point, there. But if the authorities were investigating, they'd check railroad and truck shipments of any plastic or metal aircraft construction materials into this region, and where they were delivered. They'd check local machine shops, auto-parts shops, aviation parts dealers—and *they'd check garages*! If one of us walks up to a garage, buys a battery, and walks away carrying it on his shoulder, don't you think the garage mechanic is going to remember him, what he looked like, how tall he was, what he weighed? How often does anyone without a car buy an auto battery and carry it away on his shoulder?"

"We might 'borrow' somebody's car," Smitty mused, grinning.

"We might be caught ten minutes afterward, too," Foster objected. "The police are quite efficient at catching car thieves."

"Then we need a car," Morrow concluded. "Smitty, can we lift out of here once we've rebuilt our jets?"

"We could travel a few hundred miles," Smitty conceded. "Not that it would get us anywhere."

Morrow grinned crookedly. "Would it get us to Westerton, New Jersey?"

It would. And the next night, it did.

The three men crouching in the control pit of the sleek, black ship looked red-eyed and haggard from fatigue and lack of sleep. They had stripped off their shoes and socks to let them dry near the ship's heater, and their damp, mud-stained coveralls were drying on their bodies. Foster had developed a wracking cough and his nose was running.

The air-speed indicator registered three hundred and sixty-eight miles per hour. Smitty stared at it, glumly. "Let's just hope it doesn't fade out on us again," he muttered.

The test of the ship's performance had been the whole purpose of their long, cross-country trip, Morrow thought wordlessly. They had made every preparation they could think of for the trip. Each had a special suit with helmet and gravitor-tank—and one additional feature: a one-man propulsion unit. They'd developed that in the workshop when they ran one of the suit's gravitors until its field had completely polarized the suit; then, when the suit was suspended high over a small wood fire, the smoke from the fire had risen up into the suit's gravitor field and twisted and swirled around to conform to the warp of that field. Knowing those twists and swirls, Foster had designed a small jet unit with air intake slots and jet-pipes which utilized the air-flow through the gravitor field.

Of course, there was one fault in this jet unit: it was designed to use the air-flow around a gravitor standing still. With the gravitor in motion, that air-flow was altered somewhat. But when Smitty had floated up in his suit with that little jet unit built into its tank, he had managed to fly around the sawmill yard at a good fifteen miles per hour. The air drag against his legs, since the gravitor made him weightless, was considerable—it flattened him out in horizontal flight and, by swinging his legs from one side to the other, he was quite capable of controlling the direction of his flight. The lift or descent of the gravitor sufficed for climbing or diving maneuvers. He'd looked like a human fish swimming in the sky.

For the ships, of course, such a jet unit wouldn't do. The ships needed jets which would work while in motion, at speeds exceeding a hundred miles an hour. Thus, they'd had to fly the ship until its gravitors completely polarized its hull. Then they had to determine the air-flow over that hull at flying speeds with flow and pressure indicators mounted on the hull. Then they had to rebuild the tail-jets to conform with their findings.

A flight half-way across the continent and back to their workshop would have served for that. But then, they had to be sure that there was no further change in the air-flow or polarization or gravitor field. For that reason, they had decided on this trip all the way across the country. It would give them a complete, thorough test of the ship.

They had even gone so far as to arm themselves for defense, in case they were forced down anywhere and someone tried to get rough with them. In a strictly legal sense, the streamlined plastic pistols they carried were not lethal weapons.

Technically, those pistols were ray-guns. They fired a beam of light.

That light came from a standard photographer's flash-bulb. It was focused into a tight, narrow beam by the pistol's barrel reflector. It wouldn't penetrate the human skin; it wouldn't even raise a blister. It was almost physically harmless. But directed at a person's face at a distance of no more than twenty feet, it would leave them totally blind for about three minutes. A simple flash-bulb delivered a nice, bright flash.

A person suddenly struck blind wasn't likely to be in any condition or mood to cause trouble.

All other preparations for the trip had been as completely thorough, as carefully planned. Yet they had made one slight error. They had forgotten to include extra batteries for the ship. In all their careful and intricate preparations, that one, simple precaution had been overlooked.

And now, because of it, Morrow wondered if the whole purpose of their trip wasn't going to be changed. They were flying to Westerton where he would borrow a car from someone he knew.

The one person in Westerton he felt he could trust more readily than others was, of course, Gwyn Davidson. And Gwyn's father had a car.

But they couldn't land their ship anywhere near town, where he could go directly to Gwyn. They would have to land some distance from town, at a spot he knew quite well, and he'd have to proceed from there. He couldn't hitch-hike into town; people knew him, would recognize him and ask questions. He'd have to fly in on his suit gravitor.

And when Gwyn saw that, he'd have some explaining to do. He wondered what she would think....

He wondered, too, at the thrilling tingle of excitement which was washing through him in waves of—of ecstasy, almost! There was no other word for it! He felt like a kid with his first toy.

The ship glided down through the cold moonlight and grounded behind a thick screen of trees, hidden near the shore of a small lake. Across the glistening, ice-covered lake, the sprawling log structures of Lakeshore Lodge loomed blackly against the snow glare. The buildings were deserted, uninhabited during the winter.

Morrow remembered it during the summer season, alive with people in bathing suits, and small boats out on the lake, and this small clearing behind the trees where they landed, where he and Gwyn had sprawled on a blanket, sunning themselves. He remembered the spot quite well.

Westerton was twenty miles away.

He was numb with the shock of the cold air and the weird experience of his flight when he approached the town.

*It felt so damned strange!* He was flying at about four hundred feet, sprawled flat with the wind blowing and buffeting over him. His head was protected by his helmet, of course, and he was only doing about fifteen miles an hour—but the weightless condition of his arms and legs made it feel as if he were battling a sixty-mile gale! And using his legs to guide his flight completed the impression: he *swam* through the air!

The yellow lights of town began to outline the streets and intersections below him. Never having seen them from the air, they were at first strange to him, unrecognizable—then he got his bearings and flew onward. Or swam. His breath was coming in labored gasps. His whole body was tensed against the cold seeping into his suit.

He searched frantically for Gwyn's house. It was after two in the morning; she'd be home asleep now.

He spotted it, flew over it, and cut his tiny jets. Then, tuning down his gravitor, he drifted gently downward until his feet crunched in the snow in the small back yard. Looking up, he saw with a start that he'd just barely missed straddling a telephone wire on his way down.

Shivering, he strode toward the house. It was a two-story, white frame structure and Gwyn's room, he believed, was on the left side upstairs. He went around to the side of the house and looked up at the windows, puzzled. Which was hers?

It wouldn't do to try to scramble in a window, anyway. Gwyn would probably let out a scream that would awaken the whole neighborhood—or her father might take a shot at him!

Better to do it the conventional way. Knock on the front door. Ring the bell.

Should he take her father into his confidence, too? Morrow decided against it—no point in stretching his luck too far.

Then he had to get Gwyn out of the house. Alone.

Morrow shook his head, grinning wryly. This was getting more like a kid's game all the time! Then he shuddered. It was cold as blazes! He had to get inside and get warm!

He strode purposefully around front, went up on the porch, and rang the bell. A good, long ring. Then he jumped off the porch and ran back to the side of the house.

A light flashed on upstairs. A shapely, feminine silhouette passed across the curtains as Gwyn crossed the room, pulling on her housecoat.

Morrow stepped close to the wall, tuned up his gravitor, and rose easily up to the window. He grabbed the sill to stop himself and peered in. The room was empty. The window was raised slightly.

He pushed it up, scrambled in, and lowered it behind him. The room was small and neat, littered with feminine knick-knacks, and smelling more clean and polished than sweetly perfumed. He strode past the rumpled bed and sat down in the chair against the wall, out of sight from the doorway.

His gravitor tank kept him well-forward on the edge of the chair. His suit remained ice-cold and snug in the room's warmth, which he felt seeping in through the vents in his helmet collar. He shuddered violently, then sucked the wonderfully warm air into his lungs. He gazed around, noting that his helmet gave everything in the room a bluish tint, but he was so accustomed to that he didn't mind it. Then he saw himself in the dressing-table mirror, across the room, and almost doubled over with silent laughter.

What a strange creature he was, with a shimmering, bright skin and a huge, dark globe of a head!

Gwyn would scream her lungs out!

He reached up hastily, broke the clamps on his helmet and swung it back. Best to let her see his face, first, and recognize him—

A door opened out in the hallway.

"Who is it, Gwyn?" Old man Davidson's voice had the mellowness of a concrete mixer.

"Nobody, Dad!" Gwyn's voice came from downstairs, puzzled. Small feet stamped on the stairs. "It's awfully cold out for anyone to be playing pranks. When I opened the door, there was nobody out there!"

"Well, go back to sleep, honey."

"All right. 'Night, Dad."

The door closed in the hallway. The small footsteps trod disconsolately toward Gwyn's door.

Then she was swirling into the room, closing the door, and pulling the housecoat off over her blue, pink-flowered pajamas.

When she saw him, she froze and sucked in her breath.

*"Bill!"*

It wasn't a loud exclamation, but a faint, weak cry. Morrow had his finger over his lips, motioning her to silence.

Her face went blank; then she tugged her housecoat frantically back on and strode over to him. Her voice was a low, insistent murmur. "Bill, how did *you* get in here? What *is* this, anyway?" Her wide eyes were sweeping over him from head to foot, unbelievingly. "What on earth's *happened*?"

"Sit down," Morrow said gently. "Keep your voice low. Can't let anyone know I'm here, Gwyn—and I need your help!"

Gwyn looked at him steadily for a long moment. Then she said, with a kind of silent protest, "All right, Bill. I'll get Dad's car out and go with you. Now—how are you going to get out of here?"

"Same way I got in," he told her, quietly. "I'll meet you outside."

Then, before she could protest, he strode to the window, raised it, climbed out, and shoved free—using his gravitor, of course, as he did.

She stared at him from the window until he touched ground. Then he waved to her and went around the house to the garage.

She came out a few minutes later, dressed in a warm, woolen suit.

Morrow explained the project to her as they drove downtown. When they got out on the highway, approaching an all-night garage, she dropped him off. A half-hour later, she was back.

"Got the batteries?" he asked, piling into the front seat beside her.

"Yes, I got them," she said.

They drove on out to Lakeshore Lodge.

She was grimly silent all the way. No questions, no comments whatsoever. She kept her eyes straight ahead on the highway, her face expressionless and a little pale in the passing lights.

*She doesn't like it*, Morrow thought bitterly.

But if she didn't like it, why didn't she say so? Did she think this female silent treatment would work on him? Gwyn should know him well enough to realize that such typically feminine maneuvers always have the opposite effect of what they were supposed to have on him. Silent disapproval, huh? Then the devil with her!

But such obvious deceit wasn't like Gwyn, either, he realized. Maybe it was something else, then.

Maybe she had gotten the idea that he didn't want her opinion. Suppose she wasn't asking questions because she thought he didn't want her to ask anything!

Possible, he thought. Even probable. He might have overdone it when he tried to impress her with the need for absolute secrecy. Maybe she thought he'd merely come to her because he needed help, that she wasn't included in the project itself—

But *was* she?

Morrow realized, then, that he wanted her to come back with him. Back to California, to the workshop—

What would the others say about that?

And did he want to expose Gwyn to the sort of risks they were taking?

They drove up to the Lodge and parked. "I'll have to take the batteries in one at a time, I guess," he said dourly.

"Where?" She seemed to rouse herself out of her own thoughts.

Morrow pointed across the lake. "The ship's over there, beyond the trees. Remember the place?"

"Oh!" she exclaimed softly. He couldn't see her face in the darkness.

"I'd better call them," he said, opening the car door. He stepped out into the snow, straightened up beside the car, and swung his helmet over his head. There was a tiny, pocket-sized walkie-talkie built into the helmet collar under his chin; he flipped its switch and waited for the set to warm up.

Then he began calling quietly. "Angel One to Cloud Two. Angel One to Cloud Two—do you hear me? Come in. Over."

"*Cloud Two to Angel One,*" Smitty's voice was a tiny, metallic sound inside the helmet. "*Hear you faint but clear. Give your position, over.*"

"I'm at the Lodge," Morrow replied. Gwyn was watching him, wide-eyed. "The girl is with me. We've got the stuff. I'll have to bring it one at a time to you, over!"

"*Angel One, are you observed? Repeat, are you observed? Over.*"

Morrow scowled in puzzlement. "Nobody here but us chickens," he quipped back. "What're you driving at, over?"

"*Do not attempt to bring stuff here,*" Smitty's voice taunted him. "*You might drop something. Remain at your position—we'll come there!*"

Morrow's mouth went slack. Of course! He should've thought—

"*Cloud Two to Angel One! Acknowledge, please. Over.*"

"Okay, guys!" he snapped. "Roger, wilco, over and out!" He switched off the set, angrily.

But what was he angry about?

He wasn't sure. Something was wrong, somewhere. Somehow, things just weren't working out right.

"They're—coming here?" Gwyn asked hesitantly.

"Sure," he retorted, his tones unnecessarily brusque. "They're coming here."

"Oh." She gripped the steering wheel and stared ahead, not looking at him.

"Gwyn—" Morrow started around the car, around to her side to open the door and lift her out—

A faint, whining sound came from above as he reached the front of the car. He stopped and looked up, startled.

The sleek, black ship settled down to the white snow before them. A sort of strangling gasp came from Gwyn, then she was out of the car and standing beside him, clutching his arm tightly.

The thick door swung open on the faintly gleaming hull. A figure in bright, snug garments, with a dark globe of a giant head, floated out of the door and came gliding toward them. It swung its legs down and settled to a crouch in the snow in front of them.

"Well?" the strange, dark globe-head drawled. "Don't I rate an introduction?"

The batteries were installed. The old ones they replaced were stored on the cargo deck to be recharged when the ship had returned home.

The forward lounge was bright, warm, and cheerful, with the ultra-modern interior fittings and deep, foam-rubber chairs and the moonlit snow and trees outside the long port-holes' slits. Gwyn sat between Smitty and Morrow, holding her cup out for Foster to pour her coffee. Foster poured with a deft flourish. He had his jacket tied around his waist as an apron.

"I've always maintained," he observed with mock seriousness, "that the woman's touch is absolutely essential to the success of any project attempted by man!"

"Quite true," Smitty agreed, going along with the gag. "Though I'm not a lace-curtains man, mind you. Just lace." He grinned wolfishly at Gwyn.

"Being a married man, myself," Foster went on, pouring himself a cup of coffee, "I have so accustomed my tastes to minor discrepancies as practiced by the fairer sex that I'm no longer disturbed by such. Nylon stockings and underthings hanging all over the bathroom, for example. As one gets used to that sort of thing—"

"Hear, hear!" Smitty chanted.

Foster sprawled in a dignified pose in the chair facing them. "As one gets used to it," he continued unmindfully, "it fades to its proper insignificance. Then a man can truly visualize the worth of feminine companionship—the slippers, the evening paper, the scratching of one's back—"

Gwyn was laughing. The tension was going out of her shapely, young body. Her gaze was mirthful, speculative—especially when her glance slid over to Morrow.

"One finds," Foster went on, "that the prime essence of—of—"

He broke off with a violent sneeze.

Morrow finished his coffee, set his cup aside, and rose. "We'd better take off," he said flatly. He turned and faced them.

Smitty and Foster were looking at him with a silent reproof. Gwyn's eyes were on the floor. She set her cup aside, untouched.

Morrow returned their look without expression. Something danced and giggled and rolled, hugging its sides with laughter, inside him, but he kept it off his face.

"Gwyn!" he said. His tone was sharp, insistent.

She stood up uncertainly. "I'd—I'd better be getting home, too," she said.

"Right." He nodded. "We've got to get off before sunrise catches us— we'll be safe over the Pennsylvania brush country."

"All right." She moved toward him, toward the bulk-head door at his back.

He reached out and touched her shoulder, stopping her before him. "When we get back, I'll write you," he said gruffly. "Meanwhile, you can be straightening out your affairs here, and—in a couple of weeks or so—"

She looked at him, then. Eyes wide open and shining, lips parted.

"Well, don't just *stand* there!" Smitty bellowed indignantly. "Go on and *kiss her*!"

It was hardly a month later when Morrow stood in the doorway of the sawmill-workshop, his arm around Gwyn, and said, "We need a good mechanical engineer! Can't get anywhere without him—"

And a small, gray-haired man sat up in bed, a few nights later, and stared at the two strange creatures standing before him. Their heads were dark, featureless globes. Their bodies were covered with a bright, shimmering skin. He noted vaguely that the female of the species was stacked quite well.

"Can't do anything without a good structural engineer!" the little man snapped angrily, a few months later, as they were standing around a littered workbench.

A slender, middle-aged woman stepped off the bus and walked up the quiet, dark street toward her home. Then she froze, a scream stuck in her throat, as several weird creatures swarming out of the shadows....

A Northern Airlines pilot glanced out at the port wing of the giant, humming stratoliner. His mouth fell open, then he grabbed his co-pilot's shoulder and pointed out toward the wingtip.

Two sleek, fish-like little ships were flying perfect formation with the big plane, their black silhouettes outlined sharply in the warm summer moonlight.

An Air Force pilot rode his powerful, deadly jet-fighter across the desert country, thinking of the wife and children waiting for him in Los Angeles. Suddenly, he tensed, staring over the side. Far below, a black shape was outlined against the gray earth.

Quickly, the pilot radioed his flight h.q. and fired his guns, blasting their muzzle-covers away. Then he peeled over into a dive and went screaming downward. The black shape appeared on his sights, his thumb pressed the fire-button—no time to set up for auto-fire—

And then, the black shape was gone!

The wind stopped screaming around the little ship as Smitty cut its gravitors back in, halting its helpless plunge. He pointed its needle-nose up the black maw of a deep canyon and glanced upward, grinning as he thought what must be going through that jet jockey's mind. *Which way'd he go?*

Just let 'em try following a "spaceship" through one of these twisting canyons! At a jet-fighter's thousand-mile-an-hour combat speed, just let 'em try!

But, as Morrow discovered, a heliocopter could follow anywhere. It wasn't when he and Gwyn drove into Stockton to get married, but later, when they were playing follow-the-leader in silvery wonderland of clouds under a full moon. He and Gwyn wore gravitor-units strapped to their backs, with the harness incorporated into a swim-suit attire, without helmets or any other garments. It was a warm summer night filled with cool breeze that caressed their skin as they circled and skimmed over and around the bright masses of cloud.

A civilian pilot riding his little ram-jet heliocopter southward toward 'Frisco saw them gliding around the clouds at approximately the same moment Morrow caught sight of him. The 'copter gave chase. Morrow and Gwyn parted, trying to confuse the pilot, but the 'copter swung on its whirling blades and went after Gwyn. Its speed was greater than hers and it was rapidly overtaking her—the pilot jockeying it into position so its blades would strike her. Apparently, the pilot had a morbid sense of humor.

Seeing this, Morrow swung back, intercepted the chase, and swooped low under the 'copter, trying to unnerve the pilot. But the pilot merely waved at him and laughed, shouting something about *"Gonna get one of you, anyway!"* that Morrow barely heard.

He circled and dived at the 'copter again, fumbling at his belt. This time, he pulled up to the side of the 'copter's teardrop cabin, stopping himself by slamming both feet against the cabin. Startled, the pilot jerked the controls and the 'copter dipped its blades at Morrow. He had just enough time before cutting his gravitor and plunging free to fire his flash-bulb pistol directly into the pilot's face.

Checking his fall several hundred feet below, he looked up and saw the 'copter wallowing precariously on auto-controls, its pilot pressing his hands over his eyes. Gwyn came swooping downward, her dark hair billowing out behind her, and called anxiously to Morrow—when he fell, she'd thought the 'copter blades had struck him.

They lost themselves in the starry blackness before the pilot regained his sight.

That spring season, the newspapers broke out in a rash of headlines and front-page stories about ships from outer space and life on other worlds, quoting eye-witness reports and authoritative comments. By summer, the latest best-seller book was a loosely-written volume entitled THE MONSTERS ARE AMONG US!

Those fortunate members of a certain group of thirty-seven men and women broke into grins every time they heard the book mentioned. This group had laid out a collective sum of slightly over a hundred thousand dollars for the construction of a small vacation resort in the Nevada desert.

It was a rather special resort. The buildings were built cheaply, yet were designed by certain talented engineers so that their structures were considerably stronger than those of conventional buildings using costlier materials—a not too difficult feat, considering the outmoded building codes which governed most construction—and were surprisingly sleek and ultra-modern, as well.

The members of this group usually continued their work in plants and laboratories outside. Each year, when their vacation-time came up, they would rush off to a little radio repair shop in Stockton and have a quiet talk in the back room with its youthful proprietor. That night, they would drive up into the mountains to an old, abandoned sawmill, where a strange ship would drop out of the darkness to greet them....

It was a deep, twisting canyon east of the Kawich Range. Sand-stone cliffs towered up nearly three hundred feet on each side and a spring-fed stream trickled along the boulder-strewn floor, curling around clumps of stunted pine trees and dense brush. The wind sometimes tore through the canyon with a deep, mournful whistle.

Farther up, the canyon widened out. A pile of giant boulders formed an island in the middle of the floor and cliff-dwellers had built their dwellings in a large cave half-way up on one wall. Those dwellings were now occupied and joined by slender spans to the three sleek towers rising up from the island. At the foot of the island, a flat space had been cleared and long, low sheds built around it.

In the middle tower overlooking the clearing, which was now occupied by a slender, black ship. Morrow sat before the observation wall of his living room and gazed downward. He wore a simple pair of trunks on his tanned body, and socks and sneakers on his feet.

The man sitting in the chair next to him was tall, broad-shouldered, and husky. There was a two-day growth of beard on the lean face and the soiled white trousers and shirt looked as though they had been slept in. The man's eyes were cautious and tense when he glanced over his shoulder.

Smitty was standing behind his chair. Smitty wore the same casual attire that Morrow did, with the addition of a cartridge belt and holstered pistol about his thighs. The brown hand resting on the pistol-butt—it was a Colt .45 revolver—gave their visitor silent confirmation to the fact that he was, essentially, their prisoner.

"So it took you just six months to find us, did it?" Morrow asked musingly. "Too bad about the shipping records on those plastic construction materials—you must have traced down the shipments from every company in the country before you found that."

"We traced nearly all of them," the visitor conceded. "In fact, this one would've escaped our notice if you'd used any half-reasonable company in Stockton to cover up your use of the materials."

Morrow took cigarettes and matches from the pocket of his trunks and proceeded to light up, calmly. It was nearing sunset and the canyon was already plunged into a blue twilight, in which the lights in the towers and on the small landing field below glowed softly, in soothing pastel colors.

The visitor sat unmoved through the silence. He had been caught inside the old sawmill and flown to the hidden base the night before. His credentials said he was an agent of the United States Bureau of Internal Security, that his name was David Lyle. Morrow glanced at him, speculatively.

"I've told you all I dare about our group, here," he said. "I've told you some of the things we've done—"

"Without explaining them," Lyle interjected wryly.

Morrow smiled. "You wouldn't grasp the technical end of it if I had told you. It's as if I were the first man to invent the wheel and had gathered a few others about me who were now developing the propellor, the fly-wheel, gear-ratios and the piston engine. We can generate enough electrical power in this canyon site to light a large metropolitan city, and we're now working on a means of using broadcast power and perhaps harnessing atomic energy. We already suspect some of the chemical and medical possibilities inherent in gravitor-field conditions—"

"And you have the answer to interplanetary travel at your fingertips!" Lyle muttered dourly.

"Yes, but without the financial means to do it," Morrow agreed. "Interplanetary travel won't be important for another hundred years anyway—if it is at all—since it will take that long for the world's population to reach any dangerous numbers."

"What's that got to do with it?"

"Mankind is due to reach the stage of population where he can no longer feed himself on Earth," Morrow explained. "He simply won't be able to raise enough food on this one planet to feed such numbers. Either that, or there'll be three or four atomic wars in the next few generations—if there's one, there'll be several wars—and population will cease to be a problem.

"There's been some talk of birth-control as the only logical answer to this overpopulation. It may be used, but I doubt its logic. You'll have to tell some people they simply can't have children, and on a world-wide scale you're going to have many cases where they disregard authority and have children anyway. Then, to make your authority stick, you'll have to take those unauthorized children away from their parents and kill them. You'll need a world dictatorship to do that.

"The only answer that's really logical is when this world gets too small to support mankind, go out and settle a couple more. That's where interplanetary travel becomes important, and not before. The astronomers claim there is very little likelihood of any native species of intelligent beings living on either Mars or Venus. I only hope they're right!

"But that isn't answering our present problem, is it?" Morrow grinned reflectively. "We could kill you, Mr. Lyle, but that would gain us nothing. There would be other agents following you. Also, it doesn't sit well with our attitude."

"Just what *is* your 'attitude' as you call it?" Lyle demanded.

Morrow glanced at him through narrowed eyes and replied, "Just what would *your* attitude be if you were in our position, Mr. Lyle?"

Later, as Morrow sat alone, Gwyn came out of the kitchen and joined him, perching herself on the arm of his chair.

"It'll work out all right, Bill," she murmured soothingly, running her fingers through his hair. "Don't worry about it."

Morrow shook his head. "We've got to let him go, Gwyn. We can't hold him."

"Then let's just face it," she replied, using her practical feminine approach. "The government is going to learn about our project. What can they do about it? Can they throw us into prison and confiscate all we have here? What'll they do with it? Without us, they won't understand it!"

"How much *will* they understand, I wonder!" Morrow said dubiously. "Will they realize this could ignite the present world tension into a raging war?"

Gwyn looked out on the silent, brooding canyon. "Would it, Bill? I mean—I'm not doubting you, darling—but are you sure?"

Morrow sighed wearily. "No," he said. "Not sure. I'd just rather not risk it."

"Well, if it happens, it won't be our fault." Gwyn slipped her arms around him and settled down in his lap. "Don't worry, Bill—"

It was nearly midnight when Morrow stood down on the field, with the gleaming, black ship looming beside him, and watched Smitty and Lyle, the agent, walking out toward him.

"Finished your inspection, Lyle?" he called out, his voice sharp, brittle.

"Yah. I've finished." Lyle strode up with a thoughtful expression creasing his forehead. "You got quite a lay-out here."

"Thanks." Morrow hooked his thumb at the ship's open hatchway. "Climb in, Boy Scout. We're taking you back to Uncle."

"Ah-hmmm—just a sec, Morrow." Lyle paused, lighting a cigarette. "I've been thinking about that question you asked me—what my attitude would be in your place."

"Yes?" Morrow stiffened warily.

Lyle grinned. "One of the things that surprises me is that of all the people in your group, none has spilled the beans. How come nobody talked?"

"If you had what we've got, would you talk about it?"

Lyle chuckled, flicking the ash from his cigarette. "We're back to attitudes, then—right?" He looked up, his gaze suddenly intent. "I think I've got an answer to your question now, Morrow."

Morrow squinted at him. "What're you getting at, Lyle?"

"Those aircraft construction materials you had shipped to Stockton," Lyle said quietly. "Building an experimental plane without authorization is a federal offense. The fine's five hundred dollars. You got five hundred bucks, Morrow?"

"I think so," Morrow replied cautiously.

"And you got a couple aeronautical engineers here who could whip up some kind of little airplane, haven't you?"

"Suppose I have?"

"Well, whip up something! Just so it'll get off the ground—put a motorcycle engine in it—and the Civil Aeronautics boys will have something to take their hatchets to. Plant it out at that sawmill of yours." Lyle's sombre eyes were laughing silently.

"So I'll pay a five-hundred-dollar fine?" Morrow asked perplexedly.

"Uh huh. And I can write a report that'll close this case."

"You—" Morrow broke off, staring at the calm, good-natured agent.

"The stuff you've got here is poison to today's world," Lyle said quietly. "Maybe, in time, guys like me can change all that. Until we do—" He left the rest unsaid.

Morrow let his breath out slowly. Then he extended his hand. The young agent's grasp was firm, decisive.

"If you two're through yakking," Smitty growled, shoving past them, "let's get outta here!" He mounted to the ship's hatch.

The two men followed him and the hatch folded shut, flush with the sleek hull. Then, gravs humming, the black ship lifted from the field.

It dwindled rapidly into the upper darkness.

# Table of Content

**Chapter 1**

# Introduction to Interdisciplinary Implant Dentistry

**Dr. Prof. Pranav Parashar**

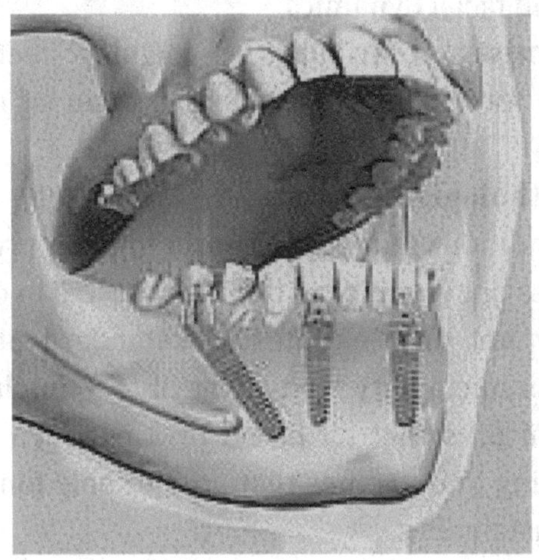

**The Evolution of Implant Dentistry Through a Multi-Specialty Approach**

Implant dentistry has undergone a remarkable transformation since its inception, evolving into a dynamic and holistic field that thrives on the integration of multiple dental specialties. In its early stages, dental implants were viewed as a highly specialized treatment, predominantly managed by oral surgeons and periodontists who focused on surgical placement and foundational support for prosthetic rehabilitation. However, as the field advanced, it became evident that achieving optimal outcomes required a more comprehensive approach, incorporating expertise from various disciplines such as prosthodontics, orthodontics, oral pathology, and even restorative and aesthetic dentistry.

**The Shift to Interdisciplinary Collaboration**

The recognition of the interdependence among specialties has redefined implant dentistry. For example:

- **Periodontists** play a critical role in ensuring the health of the surrounding soft tissues and bone, which are vital for implant success.
- **Prosthodontists** bring expertise in designing functional and aesthetic restorations, ensuring the final outcome aligns with the patient's oral and facial dynamics.
- **Orthodontists** contribute by aligning teeth or creating adequate space for implant placement in cases of malocclusion or crowding.
- **Oral pathologists** assist in diagnosing and managing pathological conditions that could impact implant prognosis, such as bone deficiencies or systemic conditions reflected in oral tissues.

This synergy not only elevates the standard of care but also enhances the predictability and longevity of implant treatments. Collaboration ensures that all facets of a patient's oral health are addressed, leading to outcomes that are not only functional but also aesthetic and sustainable.

**The Indian Context: A Growing Market for Implant Dentistry**

In India, the dental implant industry is witnessing unprecedented growth. Several socio-economic and cultural factors contribute to this surge:

1. **Rising Middle-Class Population:** The expanding middle class has increased disposable income, making advanced dental treatments like implants more accessible.

2. **Growing Awareness of Oral Health:** Government initiatives, dental camps, and social media campaigns have improved public awareness about the importance of oral health and the availability of advanced treatments like implants.

3. **Technological Advancements:** Indian dental practitioners are increasingly adopting cutting-edge technologies such as 3D imaging, guided implant surgery, and digital prosthetics. These innovations not only improve treatment precision but also streamline the

process, making it less intimidating for patients.

4. **Demand for Aesthetic and Functional Solutions:** With globalization and changing lifestyles, patients in India now prioritize oral aesthetics alongside functionality. Implants provide a reliable solution for tooth replacement that addresses both these needs.

### Addressing Complex Cases with Improved Outcomes

Integrating expertise from multiple specialties allows Indian practitioners to handle complex cases effectively. For instance:

- A patient with severe bone loss may benefit from the combined efforts of a periodontist and oral surgeon for bone grafting and implant placement.
- Cases of severe malocclusion requiring implant placement might involve orthodontic correction to create a favorable environment for implants.
- Full-mouth rehabilitations, especially for patients with systemic conditions like diabetes, demand careful planning involving restorative dentists, prosthodontists, and periodontists to ensure long-term success.

This interdisciplinary approach also enables cost-efficiency. By carefully planning treatment and avoiding redundancies, practitioners can reduce procedural complications and minimize costs, a crucial consideration in India where healthcare expenses are often borne directly by patients.

### The Outcome: Functionality, Aesthetics, and Affordability

Indian patients today seek treatments that strike a balance between function, appearance, and cost. Interdisciplinary implant dentistry aligns perfectly with these needs by ensuring:

- **Functionality:** Implants restore the ability to chew and speak effectively.
- **Aesthetics:** The collaboration of prosthodontists and aesthetic dentists ensures natural-looking results.
- **Affordability:** Streamlined processes and efficient use of resources help control costs, making implants accessible to a broader demographic.

## Economic Considerations in Integrating Periodontics, Prosthodontics, Oral Pathology, and Surgery

Implant dentistry has undergone a remarkable transformation since its inception, evolving into a dynamic and holistic field that thrives on the integration of multiple dental specialties. In its early stages, dental implants were viewed as a highly specialized treatment, predominantly managed by oral surgeons and periodontists who focused on surgical placement and foundational support for prosthetic rehabilitation. However, as the field advanced, it became evident that achieving optimal outcomes required a more comprehensive approach, incorporating expertise from various disciplines such as prosthodontics, orthodontics, oral pathology, and even restorative and aesthetic dentistry.

### The Shift to Interdisciplinary Collaboration

The recognition of the interdependence among specialties has redefined implant dentistry. For example:

- **Periodontists** play a critical role in ensuring the health of the surrounding soft tissues and bone, which are vital for implant success.
- **Prosthodontists** bring expertise in designing functional and aesthetic restorations, ensuring the final outcome aligns with the patient's oral and facial dynamics.
- **Orthodontists** contribute by aligning teeth or creating adequate space for implant placement in cases of malocclusion or crowding.
- **Oral pathologists** assist in diagnosing and managing pathological conditions that could impact implant prognosis, such as bone deficiencies or systemic conditions reflected in oral tissues.

This synergy not only elevates the standard of care but also enhances the predictability and longevity of implant treatments. Collaboration ensures that all facets of a patient's oral health are addressed, leading to outcomes that are not only functional but also aesthetic and sustainable.

### The Indian Context: A Growing Market for Implant Dentistry